AF689728

TRACKSIDE around BOSTON 1942-1962 with Lawson Hill

Eastbound B&M local freight just west of Waverly, July 24, 1951.

by Carl R. Byron

Copyright © 2000
Morning Sun Books, Inc.
All rights reserved. This book may not be reproduced in part or in whole without written permission from the publisher, except in the case of brief quotations or reproductions of the cover for the purposes of review.

MORNING SUN BOOKS INC.
ROBERT J. YANOSEY, President
To access our full library *In Color* visit us at
www.morningsunbooks.com

Published by
Morning Sun Books, Inc.
9 Pheasant Lane
Scotch Plains, NJ 07076

Library of Congress
Catalog Card No. 00-132093

First Printing
ISBN 1-58248-040-0

Color separation and printing by
The Kutztown Publishing Co., Inc.
Kutztown, Pennsylvania

THE KUTZTOWN PUBLISHING CO., INC.

ACKNOWLEDGEMENTS

This project could not have materialized without the faith and encouragement of Morning Sun Books Publisher Robert J. Yanosey and the skill of Trackside Photographer Lawson K. Hill. A special note of appreciation goes to Kevin Farrell and the Boston Chapter, National Railroad Historical Society for the loan of these images from their Archives. Each and all photos in this book are thus credited: Lawson Hill, Boston Chapter, NRHS Collection. Within these pages, it has been my honor and great pleasure to breathe life once again into long past scenes of New England Railroading.

Proofreading and review came from B&A enthusiasts Robert A. Buck, Paul T. Carver, and William S. Parker; retired B&M train dispatcher and B&MRRHS* member Preston S. Johnson; and NHRHTA** members Arthur E. Mitchell, and J.W. (Jack) Swanberg.

The EMC *Flying Yankee* cover is courtesy of Max Knoecklein.

Thank you gentlemen. Any and all errors are, of course, mine alone.

This book could not have been created without the effort of the tens of thousands of men and women who for nearly a century and three-quarters were the B&A, B&M and NYNH&H work force, and/or those of their predecessor and successor lines. Therefore, it is dedicated to them all - living and deceased - who helped make New England a better place to live and work thanks to 170 years of New England Railroading.

A final note of appreciation goes to my wife Patricia, for her indulgence as this project swirled about our home for months, greatly trying her patience.

CARL R. BYRON

* BMRRHS: Boston & Maine RR Historical Society, Inc.
**NHRHTA: New Haven Railroad Historical & Technical Association

TABLE OF CONTENTS

*Like most who are enjoying this book,
Lawson Hill was a lifelong railfan.
He also was a highly private man, necessitating this
brief and very general biography.*

BORN CIRCA 1911 AND RAISED ON THE FAMILY homestead in Waltham, MA, Lawson attended school - likely private - in Newton, graduating in 1929. Shortly thereafter he began a career in the traffic department of the Sears & Roebuck store and warehouse in nearby Brookline. After his father passed away, he and his aging mother went to live with her sister in Waban, a neighborhood in the Boston suburb of Newton. In later years he inherited the house, and lived there until his death on March 30, 1998 at age 87.

In mid-life a medical problem arose, and Lawson sought treatment in Montreal, Quebec, Canada. There he became involved with his nurse, Edith Sturgeon, who soon immigrated to Massachusetts, and married Lawson. They had no children, and Edith passed away on December 3, 1995 also at age 87!

From the early 1930's on, Lawson pointed his camera toward the Boston area rail scene. He joined the Boston Chapter, National Railway Historical Society in 1940 - one year after its founding. During the subsequent war years, he was the 'home front correspondent' to numerous NRHS members doing battle overseas.

During the Depression and War years he honed his skills as a black & white photographer of railroad passenger equipment. Many of these images remain in the Boston Chapter, NRHS collection.

After the end of WWII, Lawson began to use color slide film. Unfortunately, he - like many others - occasionally succumbed to the temptation of 'high speed' Ansco film circa 1950, and many irreplaceable shots have been lost in a red haze. Fortunately, this era of computer color correction has been able to restore several of these images to their true, magnificent colors. (See bottom image on page 116 {NH PA set in orange}).

Lawson was a fixture on the Railroad Enthusiasts Fan Trips from the 40's through the 70's. He is well remembered as a rider, slide taker and slide swapper, but no one seems to have been personally close to him, hence the limited biography.

These images of glorious post war New England railroading at its best are Lawson's legacy. Through his efforts we once again can take a B&M Pacific powered local to Fitchburg, the B&A's premier NEW ENGLAND STATES to Chicago, or a leisurely Budd RDC trip down the New Haven to Cape Cod.

Lawson gives the "Where's that train!" look at Newtonville station May 2, 1964.

Lawson and Edith Hill, July 4, 1967.

Come along and enjoy Lawson's photography as he takes us to yesteryear New England via *Trackside Around Boston 1942-1962.*

TRACKSIDE *around* BOSTON

FROM THE BEGINNING, BOSTON AND ITS ENVIRONS WERE WEDDED TO THE ATLANTIC OCEAN. After all, many of Boston's citizens or their ancestry had crossed it from Europe. By the late 18th Century, Clipper Ship trade to the Orient had made millionaires of ship captains and investors alike. Coastal Massachusetts sea trade flourished well into the 19th century.

Sea trade profits allowed Boston to proclaim itself 'Hub of the Universe'. However, inland mills and other industrial developments of the early 19th Century heralded the canal building era. The Blackstone canal linked Worcester to Providence, RI.; the Chesapeake Canal headed inland from Baltimore, MD; and finally in 1825 the majestic Erie Canal connected Buffalo, Rochester, Syracuse, and Utica with Albany, and-via the Hudson River - New York City. The 1820's found Boston justifiably concerned about losing its commercial superiority.

In 1825 the Massachusetts Great and General Court (legislature) commissioned civil engineer Loammi Baldwin - who bred the Baldwin apple - to survey a canal route 'to the west' - then presumed to be the Hudson River. His 1826 report favored a route through the northern half of Massachusetts - via Fitchburg, Greenfield, and North Adams. The only major problem was the 5 - mile wide, half-mile high Hoosac Mountain east of North Adams. Otherwise the route was *easy*. The shaken legislators paid Baldwin and tabled (killed) the report. However, Representative Abner Phelps suggested a *rail-road* in the canal's place.

Phelps was impressed by the gravity Granite Rail-Road, which on October 7, 1826 had started hauling large granite blocks from the Quincy quarry down to the waterfront for barging to Charlestown to build the Bunker Hill Monument. It is considered the first incorporated railroad in America.

By 1830 canal rage was supplanted by railroad fever, of which New England was destined to catch a severe case. During 1830 and 1831, the General Court issued corporate charters for the Boston & Providence, Boston and Worcester, and Boston & Lowell railroads. Like spokes from a wheel hub, the B&P went southwest; the B&W west, and the B&L northwest from Boston. The B&M's (Massachusetts) corporate predecessor, the Andover & Wilmington, was chartered in 1833. By now, if they could have built on the waters of the harbor, a line would be abuilding east toward Spain!

Boston immediately became the Hub of New England railroading. Late 1835 found the B&P, B&W, and B&L roads in operation, and no shortage of promoters for additional ones, or extensions. After all, it was an economic death sentence if the tracks bypassed your town. The next major Boston road, the Eastern RR, headed northeast from (East) Boston to Salem in 1838.

Accidents arrived too. On June 19, 1836 New England's first head-on collision, or 'cornfield meet', occurred on the Boston and Providence, in Roxbury, MA. Fortunately, there were no fatalities. Those arrived the following June 16, with a derailment on the B&W in Worcester. Two people lost their lives. Not until 1847 did B&M predecessor Vermont & Massachusetts - the line west of Fitchburg to Greenfield - have a deadly accident in Athol. The Eastern joined the list the following year with a head-on collision in Salem.

The Hartford & New Haven RR opened in 1839, and merged with the Hartford & Springfield in 1847 - creating the basis of the New Haven system. In 1841 the Western RR opened across the Berkshire Hills to connect the Boston & Worcester to Albany, while the Concord RR connected the Boston & Lowell up the Merrimack River valley to central/northern New Hampshire. The second major east-west railroad, the Fitchburg, was chartered in 1842, and began building west along Baldwin's 1826 survey.

On March 5, 1845 the Fitchburg opened, and later that year on November 10th, the Old Colony line from Boston to Fall River did likewise. In 1847 'Boat Train' service commenced between Boston and Fall River's Long Island Sound steamships. It continued in the grand manner for ninety years.

A fabulous mile-a-minute dash occurred in 1848 on the B&M's new line from Boston to Lawrence, the first recorded railroad publicity stunt in New England.

Another decade found dozens of short roads operating end-to-end which ultimately would meld into the New Haven, Boston and Albany, and Boston and Maine systems. The B&M and Eastern roads intensified their feud for coastal New Hampshire and downeast Maine dominance. The same occurred between the B&M and B&L for New Hampshire and the mill cities of Lowell and Lawrence. The Fitchburg and the Vermont & Massachusetts railroads gave joint service to northern Massachusetts as far west as Greenfield. Beyond there, the Troy & Greenfield and its Hoosac Tunnel were just a blip on the political and financial horizon.

With a brief respite during the heights of the Civil War (1861-65), New England's railroad fever burned on. In 1867 the Boston & Worcester and Western railroads became the Boston & Albany RR - some 25 years after their tracks first met! The late 1860's found trackage of the Boston, Hartford, & Erie arriving in Boston via Dedham. 1875 also found the Hoosac Tunnel finally opened through 4.75 miles of Berkshire granite, and a second east-west line stretched from Boston to the Hudson River valley.

In 1884 the Fitchburg took control of the Hoosac Tunnel Route, while the Boston & Maine's lease of arch-rival Boston & Lowell in 1887 left New Hampshire dominated by the B&M. The following year the Boston & Providence was leased by the Old Colony.

The post-Civil War years found Wall Street stock manipulators Daniel Drew, Jim Fisk, and Jay Gould forcing railroad consolidations for financial gain. In the process they instigated both the 1868 'Erie Wars' and the Panic (Depression) of 1873.

By the mid-1880's through the Gilded Age, J.P. Morgan and his Wall Street brokerage firm was arguably the most powerful financial institution in the United States. Morgan believed that monopoly and 'spheres of influence' were the logical tools to stabilize the railroad industry and its markets. Since the 1880's the House of Morgan had been buying up railroad securities and was capable of imposing its will - at least upon occasion - on the likes of the Vanderbilt (NYC system) and Harriman (UP/SP system) families.

At first, Morgan simply bought blocks of stock or bonds to acquire control of various New England railroads and trolley lines. By 1893 Morgan's power was such that he dictated to the B&M that its sphere of influence was north of the B&A; and the New Haven's, south. In essence the B&A Railroad became a 180-mile stub-end branch line to Boston, with no major opportunity for expansion. That same year the Old Colony (and its sub-leased Boston & Providence) went into the New Haven's fold. Within two years the Boston, Hartford, and Erie trackage - by now part of the New York & New England - likewise became New Haven property. The New Haven neared its zenith.

The 19th Century ended with the new South Station opening serving the B&A, New Haven, & Old Colony. It immediately replaced Boston's North Station as the busiest passenger terminal in America.

A Brief overview of 175 years of Boston area railroading.*

The New Century found the Boston & Albany leased to the New York Central System - with Morgan's blessing. In reaction to this move, later that year the Fitchburg RR merged itself into the Boston & Maine. New England's three largest roads were nearly complete.

In October, 1903, J.P. Morgan placed Charles S. Mellen in control of the New Haven Railroad. Mellen promptly used its considerable assets to put the B&M, Maine Central, and hundreds of local trolley lines under Morgan stock or voting trust control. By 1910 Morgan interests controlled New England railroads, trolley lines, and Long Island Sound steamer companies.

By 1912 the hue and cry against monopolistic control of New England's transportation network brought the US Supreme Court and the House of Morgan, into confrontation. In 1913 J.P. Morgan died. By 1914 his firm - under Court Order - was forced to divest itself of controlling interest in all these New England transportation industries, leaving many including the B&M and New Haven lines in near bankruptcy. Scandal filled the air, and hundreds of millions of dollars were unaccounted for. At the ICC hearings the rich and powerful simultaneously acquired Alzheimer memories, and misplaced ledgers. Soon enough the horrors of WWI overshadowed it all.

With America's entry into the war on April 6, 1917, the railroad industry soon found itself under the control of the United States Railway Administration (USRA). By the end of USRA control in 1920, an exhausted New England railroad network began its long descent from power and glory.

The 1920's was a period of financial reorganization and rebuilding for the B&M and New Haven; less was needed on the B&A. The New Haven, under the Edward Pearson management, avoided post WWII bankruptcy, and by 1925 reigned once again as one of America's elite passenger roads. I-4 Pacifics and EP-2 electrics carried socialites, politicians, and common folk alike over the Shore Line. Massive 2-10-2 and 4-8-2 type engines led the freights. Only local folks noticed the few line abandonments.

The B&A continued with a degree of autonomy that Bostonians thought proper, but which was viewed by Executives of other NYC controlled roads with great envy. The first of twenty 4-6-4 Hudsons arrived in 1928, to speed the Boston Section of THE 20TH CENTURY LIMITED. New 4-6-6T tank engines also arrived, for the commuter traffic via Boston's Newton Highland Branch and the Main Line. Freight now moved with speed behind fifty-five of Will Woodward's Superpower 2-8-4's from Lima Locomotive Works. The wheel arrangement soon became known as the 'Berkshire' type in homage to their conquest of Western Massachusetts' hills.

Up north, the B&M too, readjusted. Serious consolidation and branch line pruning began, and in 1928-29 B&M received 25 Lima 2-8-4 Berkshires. They called them 'Lima's' - period. Passenger service used a large fleet of Pacifics, Atlantics, and Moguls. The mid-to-late '20's presidency of George Hannauer brought major plant improvements to the B&M, including automated hump yards, major line improvements/relocations, and the massive North Station/Boston Garden complex.

The Depression, of course, affected Boston's railroads as thoroughly as everywhere else. By 1932 and 1933 business was off by half or more from 1929. The B&A remained solvent throughout the '30's, thanks in part to its New York Central Lines parent. Boston & Maine also survived the Crash financially intact, although the Edward French Management paid its last common stock dividend in 1932! The New Haven was forced into reorganization in 1935, somewhat as a legacy of its Mellen/Morgan excesses. Vice-President Howard Palmer became one of New Haven's three court-appointed Trustees, and would guide the road until 1948.

Yet staid New England was quick to respond to the lure of the Diesel. The New Haven purchased New England's first diesel - Alco's demonstrator 600 HP switcher - in 1931; the B&M followed in 1934. Also noted was the success of the new Streamliners: the Union Pacific's CITY OF SALINA, and the CB&Q's BURLINGTON ZEPHYR. Both visited Boston to huge crowds in 1934. On February 9, 1935 a near-twin to the ZEPHYR arrived on the B&M. Assigned to Boston - Portland - Bangor, ME service, the stainless steel *Flying Yankee* was an instant sellout. Ten Lima-built Heavy Pacifics rounded out the passenger fleet with assistance from 18 Baldwin-built dual-service Mountains by 1941.

Not to be outdone, New Haven soon had a contract with Goodyear-Zeppelin for a three-car aluminum speedster for mile-a-minute Boston-Providence-Boston service. The COMET hit 109 MPH in testing, and official service started June 5, 1935. Convinced that smooth-sided, lightweight passenger cars were the way to go, by 1938 New Haven had a fleet of 205 'Moderne' styled 'American Flyer' cars. EP-3 and new EP-4 electrics handled them west of New Haven, and streamlined Baldwin 4-6-4 I-5 Hudsons did the honors to Boston. All three roads purchased only new diesel switchers from this point on, as the day of the passenger and freight diesel neared.

The B&A did not get involved in the streamliner developments of the '30's. Pullman-Standard and Steam were its pre-war fashion. Indeed, the Boston Section of THE 20TH CENTURY LIMITED was eliminated on June 15, 1938, with the advent of Raymond Loewy's streamlined 1938 CENTURY. A heavyweight Pullman NEW ENGLAND STATES LIMITED replaced it. Hudson and Berkshire power still ruled the B&A main.

Throughout the '30's abandonment petitions picked up speed, especially after the '38 Hurricane' blew through that September. The New Haven's Shoreline was devastated, while the B&A through the Berkshires, and the B&M Connecticut Valley and Fitchburg mains were also heavily damaged. Little washed-out branch line trackage was replaced.

The War Years were Hell - but a profitable one - for Boston's railroads. New Haven assigned its 60 new Alco diesel-electric DL-109's in pairs to Shore Line passenger trains during the day, then to Shore Line or Springfield line freights overnight. The B&A got 4-8-2 Mohawks (NYC's term for Mountains) while some of their Hudsons went west. B&M's Fitchburg main received twenty-four EMD FT Freight A-B sets in 1943 and -44.

The post-war years looked so bright. The New Haven alone ordered two hundred and seven new streamlined cars and 27 PA's to pull them, while the B&M received 21 E7's and split an order for 24 passenger cars with the Maine Central. The B&A's flagship was the Budd-built stainless steel streamlined NEW ENGLAND STATES LIMITED behind matched EMD E-7 or Alco PA power.

The New Haven management of Howard Palmer was replaced in 1948 by that of aged financier F.C. Dumaine, Sr. Dumaine had controlled the Amoskeag woolen mills of Manchester, NH for years. Shutting them down in 1932 and turning thousands out into the street, he converted Amoskeag into a financial holding company. His Wall Street assistant in quietly acquiring a majority of NH stock was one Patrick B. McGinnis. The axe swung for nearly three years on the New Haven until the Grim Reaper's scythe claimed F.C. Dumaine on May 27, 1951. Reportedly his final words were, 'It wasn't worth it'.

The postwar B&M continued a steady course under the French Administration, with plenty of new Diesels, and cash in the bank - none of which went to the common stockholders.

The Golden Age of the Automobile arrived with the '50's. Boston's suburbs expanded exponentially and rail usage subsequently declined precipitously. Steam ended on the B&A in 1951, the New Haven in 1952; and the B&M in 1956. The Maine, New Hampshire, Massachusetts, and Connecticut turnpikes were completed during the mid- to late-'50's, and drained most intercity passenger and much freight business. Even the New Haven's legendary Shore Line Parlor car service was decimated by the Boston - New York air shuttle.

The most noteworthy Boston rail story of the '50's was the Patrick B. McGinnis saga. A Wall Street promoter, Pat had made a bundle for himself and his investors by buying up bankrupt railroad stocks and other securities at pennies - on - the - dollar. The reorganized road likely would buy them from Pat at a higher price as part of the settlement to be released from bankruptcy, and McGinnis & Co. pocketed the difference. These financial manipulations had brought McGinnis to Frederic C. Dumaine, Sr.'s attention.

(Continued on page 6)

**Note:* This history is selective, not definitive. The reader is referred to George Baker's *The Formation of New England Railroad Systems* (1937), and/or the archives of the B&MRRHS, NHRHTA, and NYCSHS organizations for additional information on the B&M, New Haven, and B&A railroads respectively.

With the death of F.C. Dumaine, Sr. in 1951, son F. C. 'Buck' Dumaine Jr. took control of the New Haven. Buck was a breath of fresh air. Money for track work, locomotives, and passenger equipment appeared, along with a public commitment to improve passenger service. With cash flowing back into equipment and maintenance, it was not available for stockholder dividends. Aroused, McGinnis 'voluntarily' led a stockholder proxy fight in April, 1954, winning by several proxies.

The New Haven McGinnis era lasted 22 months. The flamboyant Irishman promised wonders like three Talgo trains for 2.5 hr Boston - New York service, but actually employed creative financing to maximize stockholder dividends, slashed maintenance, and sold everything not nailed down. What remained was painted a flashy vermilion, white, and black scheme developed by wife Lucille's artistic associates. (Yes, she was on the payroll, too). The GE EP-5 electrics ordered by the Dumaine administration arrived in the new scheme. Lucille had on a matching outfit for the occasion. A commuter rebellion of Madison Avenue executives helped force him out - but not far enough out....

McGinnis' real dream was to merge the New Haven and Boston & Maine railroads. The Public Utilities Commissions said, NO, and so did the ICC. While he was still at the New Haven, McGinnis' associates in 1954 had wrested control of the B&M's board from the French/Sughrue management. On January 20, 1956 Pat was forced out of the New Haven's Presidency only to become B&M President some 18 hours later!

Mid-1956 found the B&M's 48 wartime EMD FT freight locomotives being 'rebuilt with some remanufactured parts' into 50 new GP-9 engines. They arrived in a bright blue, black, and white scheme clearly related to Lucille's touch on the New Haven.

Everything that McGinnis did on the New Haven he did thrice over to the B&M. Its last profitable year was 1957, and the deficits mounted - along with massive layoffs, deferred maintenance, stockholder dividend payments, and questionable land deals bordering on liquidation.

Boston to Troy passenger service over the B&M ended on January 19, 1958. Four months later the B&A quietly sold its Newton - Highland Branch to Boston's Metropolitan Transit Authority (MTA) for conversion to trolley service.

On June 30, 1959 the new Southeast Distress - oops - Expressway opened and Massachusetts terminated subsidy payments to the New Haven for its Old Colony service. The next day the New Haven abandoned all Old Colony passenger service from Boston to Cape Cod and southeastern Massachusetts. State highway planners assured Beacon Hill the services would never be missed.

The 1960's were even tougher times for Boston's trains. In 1960 the New Haven received the final 30 of a total 60-unit order of EMD FL-9's, so-called diesel-electric-electrics. They never did replace New Haven's electric catenary, yet NH #2059 is the world's last EMD F-unit. The New Haven entered its final bankruptcy in 1961, with McGinnis associate George Alpert appointed as trustee!

Patrick McGinnis resigned as B&M President in 1962 upon being promoted to the created position of Chairman of the Board. Passenger service was gone to Portland by 1965, the same year he and three associates were convicted of kickbacks on the sale of B&M streamlined passenger cars. McGinnis served an 18-month jail sentence in the US Federal Prison, Danbury, CT. during 1966-67.

On April 3, 1961, the leased Boston & Albany was formally merged into the New York Central System, marking the corporate end of the B&A. At one point in time during these years the NEW ENGLAND STATES was reduced to a single snackbar coach, and commuter service ran only to Framingham. The first half of the '60's found half (roughly) of the B&A right-of-way from Boston's circumferencial highway Route 128 to downtown sold and converted into the 8-lane Mass Turnpike Extension. The B&A's Exeter Street passenger car yard was reincarnated as the Prudential Center, with luxury condos, apartments, shops, and business suites. Buck Dumaine even moved the Amoskeag Corporation into the 52-story skyscraper!

By the mid-60's the B&M was in shambles. North Station saw as few as 4000 passengers a day, and there was serious talk of the complete abandonment of commuter service. For years Massachusetts had an annual train - off/subsidy battle on Beacon Hill (site of the state legislature). After a couple of days with larger than usual traffic jams, the subsidy would pass for yet another year. More importantly, in 1964 the MBTA - Massachusetts Bay Transportation Authority - was formed to create a regional transit authority.

On February 1, 1968 the Pennsylvania and New York Central Systems merged themselves into 'Penn Central'. It swallowed the New Haven on January 1, 1969. The decline starting at the end of the Morgan-Mellen era terminated in the New Haven becoming the stepchild in the marriage of two arch-rival railroads. Aged and feeble NYC system E-units began replacing younger but completely non-maintained NH FL-9's on Shore Line trains at South Station.

If the Sixties ended badly, the 70's started worse. Within months, Penn Central collapsed into bankruptcy - the largest corporate bankruptcy in American history up to that time. In March, 1970 the Boston & Maine also sought bankruptcy protection. The largest solvent New England railroads were now the Central Vermont, Maine Central and Bangor & Aroostook!

State and local governments began to get seriously involved in railroad operations. On May 1, 1971 that Federal Government creation, Amtrak, took over nearly all passenger train operations in the USA. The B&A main became freight only beyond Framingham, and the number of inter-city trains out of South Station declined by nearly half. The eastern half of the South Station was removed, and talk arose of its complete demolition.

The mid- and late-70's found the B&M avoiding Conrail inclusion, selling its commuter trackage and equipment to the MBTA, and even taking over the South Station ex-B&A/NH commuter line services under operating contract. New locomotives and cars arrived for both MBTA and Amtrak service. Public agencies such as Amtrak, Connecticut Department of Transportation, and the MBTA now had control - if not ownership - of the Boston-NYC corridor, and associated commuter districts.

The 1980's mark the beginning of the resurgence of Boston commuter rail from both North and South Stations. Amtrak too, began to improve, and in 1988 a freshly rebuilt South Station greeted its first passengers.

The Boston & Maine came to its end on June 30, 1983. Tim Mellon, of the Pittsburgh banking family, purchased the B&M for $26.4 million, the sum necessary to satisfy the bankruptcy Trustee and settle its outstanding debts. The B&M's corporate charter is not turned in, but its spirit is gone; destroyed by Guilford, and its 'Springfield Terminal' machinations.

The last decade has seen the continued passenger growth with the reinstitution of commuter MBTA passenger service over long abandoned lines. Two of the Old Colony lines are back in service, with one pending and others on the drawing board. Worcester Service has returned. North side service has returned to Newburyport, and may to Nashua, New Hampshire if political cooperation can be maintained. Even Boston - Portland service is scheduled to return in 2000 via Amtrak. Unfortunately for North side riders, North Station is now a memory, replaced by a wide corridor in the new Fleet Center.

The 20th Century ends with the Shore Line electrified to Boston and in public (Amtrak; CDOT; MBTA, MTA {NYCity}) ownership, giant CSX owning the old B&A, and Guilford/B&M the only New England access for CSX competitor Norfolk Southern. It's been quite a show across the past 175 years. It will continue well into the 21st Century - regardless of what names are on the buildings and equipment!

Now, let's go TRACKSIDE in the years 1942-1962 with Lawson Hill.

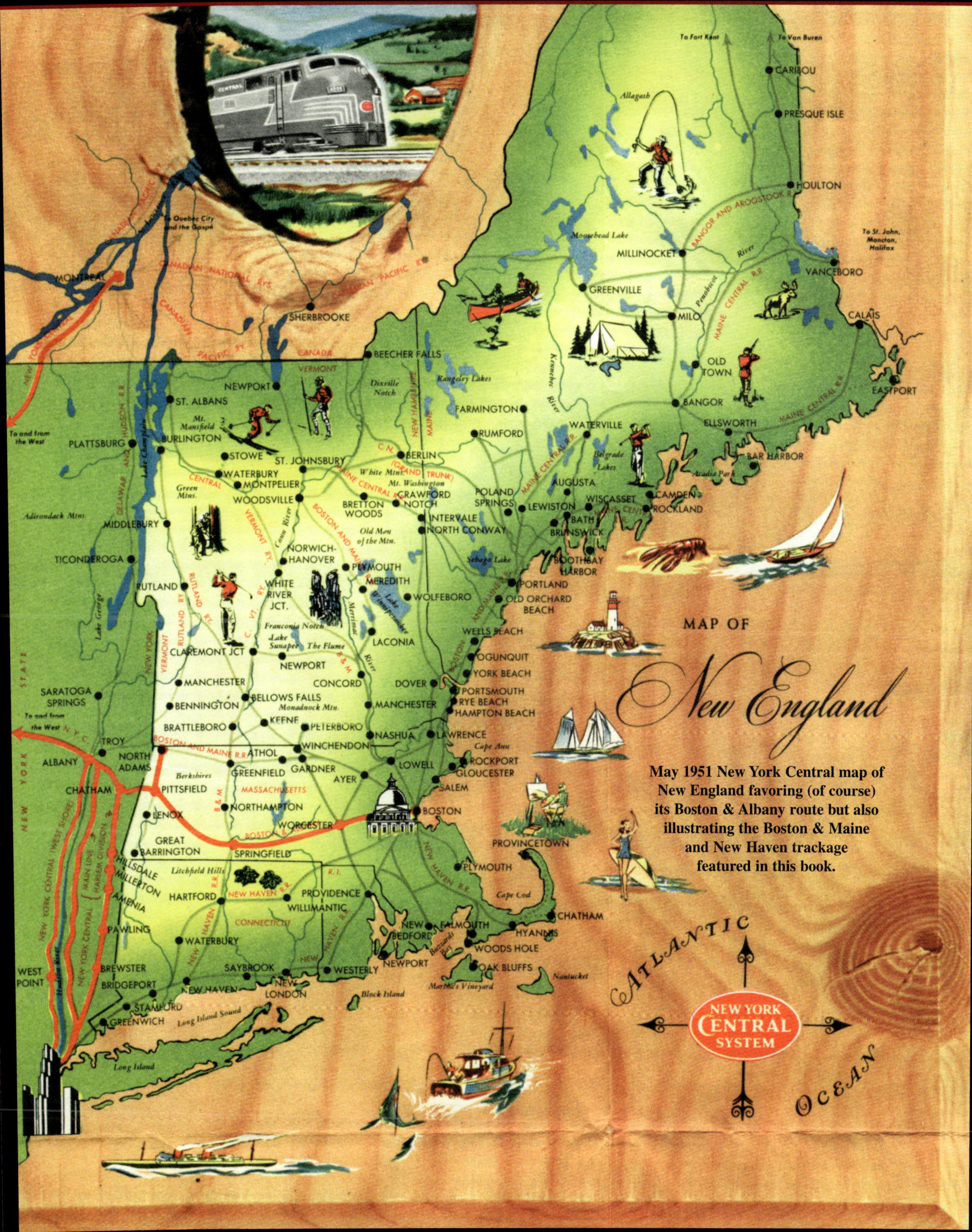

May 1951 New York Central map of New England favoring (of course) its Boston & Albany route but also illustrating the Boston & Maine and New Haven trackage featured in this book.

TRACKSIDE at LAWSON'S FRONT PORCH

Throughout his adult life Lawson Hill lived in Waban, a section of Newton. The Boston & Albany's Newton Highland Branch - now the MBTA's Riverside Line - passed through Waban. However, the B&A main line also passed through Newton with five local stops: Newton, Newtonville, West Newton, Auburndale, and Riverside. The main stop - and the LIMITED's only suburban stop - was Newtonville. There the depressed four-track right-of-way with numerous overhead bridges gave Lawson a great place to photograph his beloved trains. He captured images from four decades past in the process.

The NEW ENGLAND STATES, Train #27, stops at Newtonville before continuing its trek to Albany and Chicago. Nearly new Alco PA #4201 and its 4300-series PB-unit look great on the headend of the B&A/NYC's premier Chicago train. Note the coupler shroud still in place on this February 23, 1948.

By 1947 road diesels were making an appearance on the B&A. Here nearly-new NYC E7's #4010 - 4106 pause at Newtonville. The unidentified train is likely the inbound NEW ENGLAND STATES, judging from the streamlined consist. The new all-stainless steel 'STATES' will not arrive from Budd for 18 months from this sunny November 2, 1947.

Here She Comes! With an exhaust plume reaching 150' or more in the air, a westbound New York Central 3000-class 4-8-2 Mohawk announces its impending passage through Newtonville. Although March 12, 1950 is near the end of B&A steam, this image captures the visual power of the steam locomotive for all time.

On a sunny and clearly cold November 20, 1941 a B&A Hudson flying green flags (another section following) rolls east through Newtonville. This is Lawson's earliest color slide; he has but a handful prior to the end of WWII.

How's this for rarity? NYC's first passenger ABA F3 set - #3500-3600-3501 - in one of the NYC's earliest F-cab paint schemes. The B-unit is steam locomotive black, and the side stripe doesn't reach the end of the A-unit. The steam generators are working - note the twin white puffs at the rear of #3500.

F3A #3503 poses for a close-up on November 21, 1947. The Ansco film has tinted and faded, but still the image holds clear details. The high-shrouded fans, the steam generator exhaust, and even the fireman are crisp in definition.

Although the Diesel was new and excitingly different, it was steam that brought Lawson trackside. On a sunny but cold February 23, 1948. Lawson picked up his camera and once again visited Newtonville. From the overhead bridge we study westbound Train # 23, a Worcester local behind B&A Pacific #587. Eastbound local #550 is dropping off passengers (only - no inbound pickup) in the background. Seeing oncoming exhaust high in the sky, Lawson hurries down to track level.

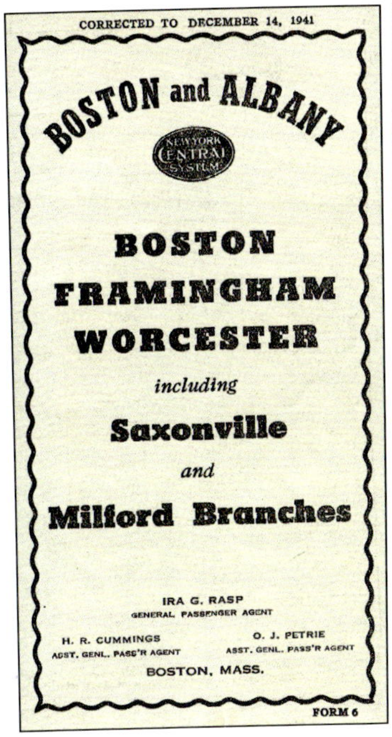
CORRECTED TO DECEMBER 14, 1941

BOSTON and ALBANY

NEW YORK CENTRAL SYSTEM

BOSTON
FRAMINGHAM
WORCESTER

including

Saxonville

and

Milford Branches

IRA G. RASP
GENERAL PASSENGER AGENT

H. R. CUMMINGS
ASST. GENL. PASS'R AGENT

O. J. PETRIE
ASST. GENL. PASS'R AGENT

BOSTON, MASS.

FORM 6

Descending to trackside, momentarily Lawson encounters B&A Train #75, a Framingham local, with Pacific #576 in charge.

Next the baggage-smoker and four coaches of Train #551 roll west behind NYC J1e Hudson #5323. If on schedule, the time is 10:31AM.

With all the glory steam had to offer, L3a Mohawk #3024 rolls THE PAUL REVERE west from Newtonville on March 12, 1950. Little more than a year remains of B&A steam.

On April 26, 1947 B&A Train # 97 - the PAUL REVERE - rolls west through Newtonville behind B&A J2c Hudson #615. There are three new streamlined cars in the Chicago-bound consist.

By the late '40's, it was apparent that the B&A was determined to replace steam power as quickly as possible. It was now-or-never for steam photography. On February 27, 1949 L3a Mohawk #3012 awaits the conductor's signal on Train #49, the Boston Section of the ADVANCE KNICKERBOCKER, at Newtonville.

Just east of the West Newton station NYC Pacific #4713 has a Framingham local well in hand on March 12, 1950 as they roll the last miles east to South Station. Note the Tydol gas station sign.

Another Interloper! B&A Train 20 - the first AM local from Springfield 'except Sundays' - heads inbound near West Newton behind an Alco RS-2 for its 8:56 AM arrival at South Station on April 8, 1950.

NYC E7 #4035 leads E8 #4057 on Chicago-bound #27 at Newtonville on June 12, 1954.

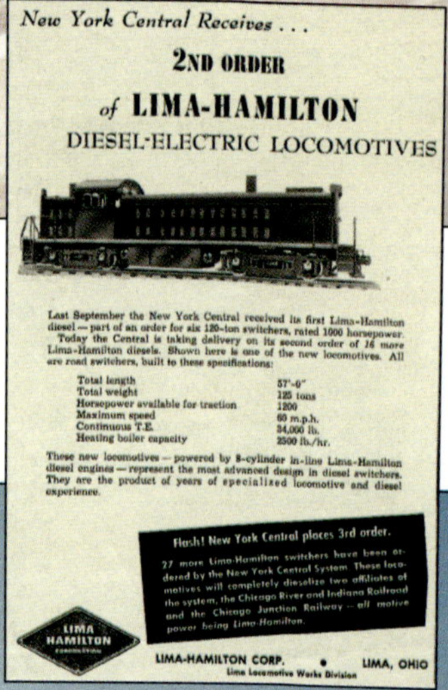

A locomotive type unique to the Central. On June 12, 1954 NYC #6214, a Lima-Hamilton 1200 horsepower road switcher, handles local #23 at Newtonville. Note the short hood steam generator stack. Unlike the 40-year reign of the 2-6-6T and twenty year reign of the 4-6-6T 'tank' engines in commuter service, the high maintenance orphan L-H units will last little more than a decade before being replaced by Alco RS-2 & -3 models.

NYC L3a Mohawk #3021 rolls Train #46, the eastbound INTERSTATE EXPRESS, through West Newton. The date is April Fool's Day, 1950.

The ADVANCE KNICKERBOCKER awaits departure in Newtonville on August 19, 1951.
The Alco RS-3 road switcher duo of #8225 and mate are a lot newer than the heavyweight Pullmans behind them.

NYC Pacific #4804 on Framingham local #519 heads into the setting sun on April 8, 1950.
The slight blur - whether accidental or not - certainly gives the impression of great urgency and purpose.

The Boston & Maine Railroad

Lawson Hill was born and raised in the town of Waltham. Located 9.9 miles west of the B&M's North Station, rail service arrived in 1845 via the Fitchburg RR. The Boston & Maine Railroad took over in 1900. Over the years, Lawson spent many pleasant days on and about the B&M from North Station to the furthest trackage of its northern New England empire.

It's hard to imagine today how much power and influence Boston's railroads had over New England's financial and transportation worlds during the first half of the 20th Century. Until the superhighways and jetliners of the 1950's, most trips were by train, and most work was in the city itself, not around it in the 'burbs'. North Station, with its 23 tracks and four lift bridges over the Charles River, was a very busy place indeed.

Even today there are four distinct - and historic - routes leaving North Station. The western most is the Fitchburg, named for the first major city west of Boston on the Troy/Rotterdam Jct., NY (Hudson River valley) line. Near milepost 136 the line enters the 4.75 mile Hoosac Tunnel, passing under the Berkshires, not over them like the B&A. Hoosac was Massachusetts' 19th Century 'Big Dig', with financial and political scandals to put the current 'Dig' to shame.

Moving eastward down the platform we reach the New Hampshire, or Lowell line. Opening in 1835 as the Boston & Lowell, it eventually became the fastest route to the major mill cities of Lowell, MA. and Manchester NH., and the New Hampshire State Capitol - Concord. North of Concord the Lakes Region and the White Mountains all had rail service to Boston via this line.

Past the Lowell line comes the Western Route to Portland-B&M's 'candy route'. The 1935 streamlined FLYING YANKEE rolled the 115 miles to Portland in 115 minutes. This is the original Boston & Maine, and passes through the mill cities of Lawrence and Haverhill, MA., through seacoast NH, and up to Maine's largest city, Portland. Service on the Maine Central connected with Northern or Coastal Maine, and the Canadian Maritime Provinces.

The eastern most side of the station is the Eastern Route main. Originally the Eastern RR, which waged a running battle with the B&M for Maine service throughout its entire Mid-19th Century existence. Running to Portland near the coastline all the way, with its only major intermediate city the sea port/ship building city of Portsmouth, NH, the line held its own until the 1950's, when it was abandoned north of Portsmouth. With Boston Brahmins' North Shore estates scattered along the Eastern route through Beverly, Wenham, Hamilton, and other posh towns, commuter patronage has always been heavy.

B&M Steam Locomotive Roster circa 1945-1956 (End of Steam)

WHEEL ARRANGEMENT	CLASS	NUMBER	BUILT	BUILDER
0-6-0	G10	#200-309	1903-1910	Alco: Manchester
	G11/a	#400-429	1911-1913	Alco: Manchester
	G11b	#430-452	1916	Alco: Brooks
	G11c	#830-32/834(exMeC)	1917-1920	Alco: Schenectady
0-8-0	H1a/2a	#600-601/610-631	1916/1922	Alco: Schen./Brooks
	H3a/b	#640-654	1927/1929	BLW
2-6-0	B15	#1360-1431;1435-1459	1903-1907	Alco: Manchester
	B15a	#1460-1499	1907-1910	Alco: Manc.& Schen
2-8-0	K7/7a	#2360-2365/2370-2389	1905/1906-07	Alco: Schenectady
	K7a	#2390-2429	1907-1911	Alco: Schenectady
	K8a	#2600-2639	1911	BLW
	K8b/b	#2640-2389; 2690-2709	1913	BLW/Alco; Schn.
	K8c	#2710-2734	1916	Alco: Brooks
4-4-2	J1/1a	#3204-3209	1902	Alco: Schenectady
	J1a	#3210-3219	1907	Alco: Schenectady
	J1b	#3220-3244	1908-1909	Alco: Manchester
4-6-2	P1	#3600-3611	1910	Alco: Schenectady
	P2a	#3620-3659	1911	Alco: Schenectady
	P2b	#3660-3679	1913	Alco: Schenectady
	P2c	#3680-3699	1916	Alco: Schenectady
	P3a	#3700-3709	1923	Alco: Schenectady
	P4a/b	#3710-14/3715-3719	1934/1937	Lima
	P5a	#3696-3699(ex DLW)	1924	Alco: Brooks
2-8-4	T1a	#4000-4019	1928	Lima
	T1b	#4020-4024	1929	Lima
4-8-2	R1a	#4100-4104	1935	BLW
	R1b	#4105-4109	1937	BLW
	R1c	#4110-4112	1939	BLW
	R1d	#4113-4117	1941	BLW

B&M all-time Diesel-Electric Locomotive Roster

NUMBER	BUILDER	TYPE	YEAR BUILT
#1-2 TALGO	FM	P12-42	1957
#110-119	GE	44-Tonner	1940-1948
#200-212*	EMD	GP38-2 (*#212 renum'ed #200 in 1973)	1972
#300-317	EMD	GP-40-2	1977-1978
#800-807	EMD	SW-8	1953
#860-865	Alco	S5 (800 HP. 1st domestic use of 251 engine)	1954
#1100	Alco/GE/IR		1935
#1103-05/1106-08	EMC	SC/SW	1936/1938
#1109-1132	EMC*	SW-1 (*EMD after 1/1/41)	1939-1953
#1101 renum'ed 1161	Alco	HH-660	1938
#1102 renum'ed 1160	Alco	HH-600	1934
#1162	Alco	HH-660	1939
#1163-1172	Alco	S1	1944-1949
#1173-1188	Alco	S3	1950-1952
#1200-1213	EMD	NW-2	1941-1949
#1220-1231	EMD	SW-9	1952-1953
#1260-1265	Alco	S2	1944-45
#1266-1273/1274	Alco	S4 (#1274 rebuilt from S3 #1174)	1950/1957
#1280-1283	Alco	S5m (S5 Upgraded to 1000HP)	1962-1966
#1500-04/1530-1534	Alco	RS-2 (Passenger/Freight)	1948-1949
#1505-19/1535-1545	Alco	RS-3 (Passenger/Freight)	1952-1955
#1550-1553	EMD	BL-2	1948
#1555-1577	EMD	GP-7	1950-1953
#1700-1749	EMD	GP-9	1956-1957
#1750-1755	EMD	GP-18	1961
#3800-3820/3821	EMD	E7/E8	1945-49/1950
#4200-4224AB	EMD	FT A/B sets	1943-1944
#4224-4226AB	EMD	F2 A/B sets (Passenger)	1946
#4227-4228AB	EMD	F3 A/B sets (Passenger)	1948
#4250-4264A	EMD	F2 A-cabs (Freight)	1946
#4265-4268A/B	EMD	F7 A-cabs/B-units (Passenger/Freight)	1949-1950
#6000 *Flying Yankee*	EMC/WINTON/BUDD	B-2-2-2	1935

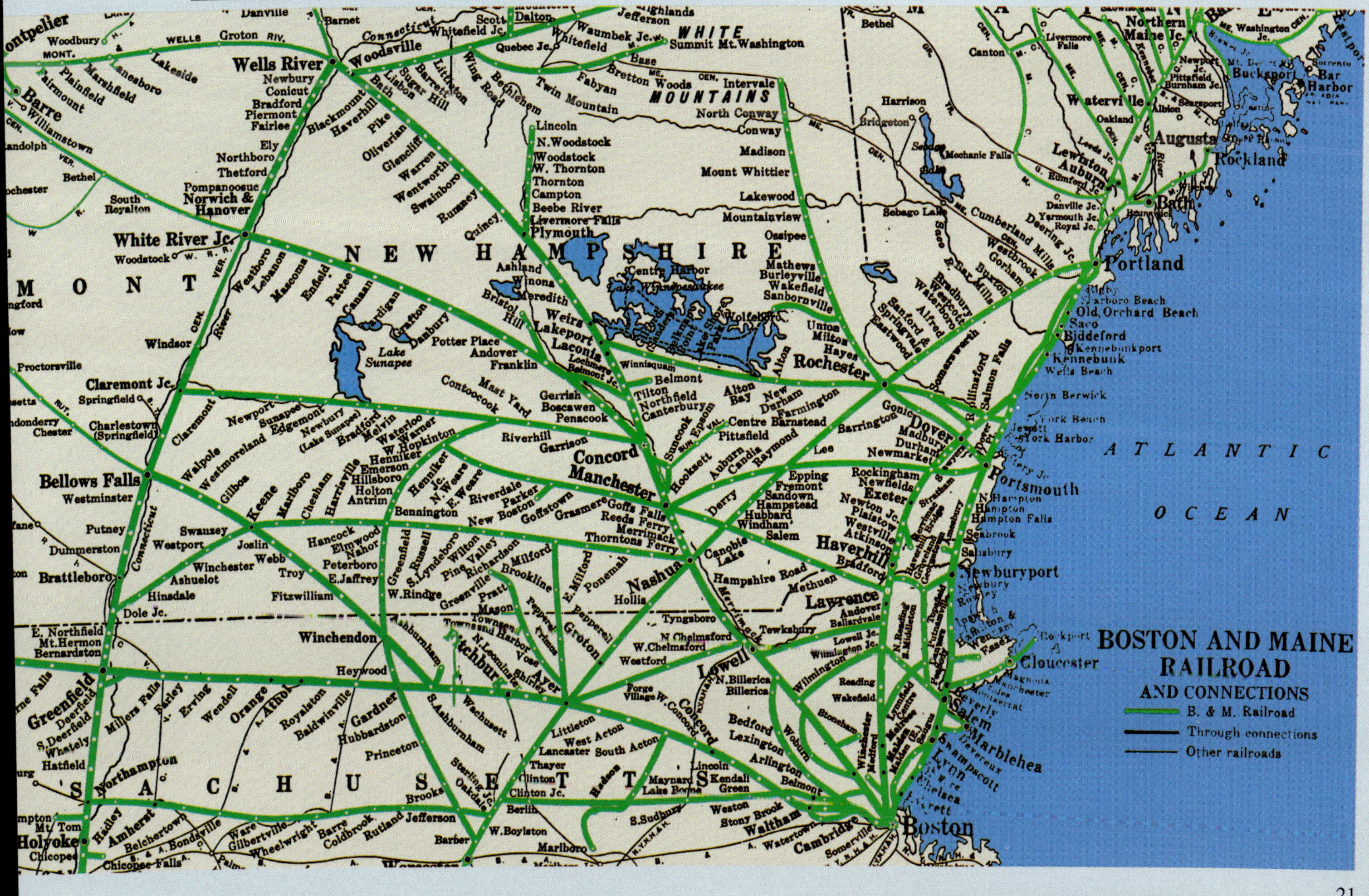

Trackside at North Station

THIS WEEKEND THE *FREEDOM TRAIN* IS IN North Station. Let's take a look!

It's a sunny Saturday, October 11, 1947. The *Freedom Train*, brainchild of the American Heritage Foundation and the US Department of Justice, has been placed for display on Track 23, adjacent to Nashua Street. The line is long for the seven-car train, whose white with red and blue striped cars have been supplied by the Pennsy. North Station looms large in the background, while to its left stands the 13-story Industrial Building - B&M Headquarters.

Inside these cars are the documents representing the founding of the United States: The Declaration of Independence, The Constitution, and Emancipation Proclamation being prime examples. US Marines, whose blue and white uniforms complement the train itself, guard all; there will be no gatecrashers at these doors. Freedom in 1947 is still a limited option to Black America, as the demand by American Heritage Foundation that all Americans be allowed to visit the train unrestricted, limits its number of stops in the still segregated South.

Alco/GE PA #1776, *The Spirit of 1776*, idles within sight of the Bunker Hill Battle Monument. (The battle was actually fought on Breeds Hill!) The 2000 hp passenger locomotive will be sold to the Gulf, Mobile, and Ohio after the completion of its patriotic duties. With the V-16 model 244 engine's habit of belching black 'Alco Dust', one wonders how many times that engine roof had to be washed while in the white paint! The trucks are painted a light gray, and the front coupler is also painted, or perhaps whitewashed.

In the grand manner, a Pullman observation complete with brass-railed platform completes the Freedom Train. While a couple of the Marine guards chat, a B&M maintenance man is catching his breath after performing some work under this car. Note the gooseneck light fixtures for evening illumination.

DRAWBRIDGES (MAIN LINE).

Draw No. 1 — Tower "A"
" No. 7 — Eastern Route
" No. 8 — Western Route

DRAWBRIDGE SIGNALS.

Draw No. 1 is protected by semi-automatic color light dwarf interlocking signals and by power operated gates controlled from Tower "A".

A warning whistle will be sounded just prior to opening drawbridge.

Draw No. 7 is protected by semi-automatic color light dwarf interlocking signals, also by gates on each side of Draw, displaying target board by day and red light by night over each track. Also by red flags by day and red lights by night placed between the rails of each track on each side of Draw. At night a green light in Draw Tower indicates Draw closed.

At Draw No. 7, train stopped by semi-automatic color light dwarf interlocking signals may proceed over Drawbridge on hand signal given by drawtender or, in his absence, must be preceded by a flagman. After passing over drawbridge, train may proceed at restricted speed as per General Rule 509-B.

At Draw No. 8, both main tracks, the Medford Track and Boulevard Track are protected by semi-automatic color light dwarf interlocking signals. Both main tracks are also protected by gates on each side of Draw displaying target boards by day and red lights by night over each track. Also all tracks over Draw are protected by red flags by day and red lights by night placed between the rails of each track on each side of Draw. At night a green light in Draw Tower indicates Draw closed.

At Draw No. 8, train stopped by semi-automatic color light dwarf interlocking signals may proceed over drawbridge on hand signal given by drawtender or in his absence must be preceded by flagman. After passing over drawbridge, train may proceed on signalled tracks at restricted speed as per General Rule 509-B.

The Draws welcome Lawson with an appropriate salute! Since the Charles River has long been a shipping channel, the draws move quite regularly. The left two draws were removed in the early 1960's. Today the ships are gone, but pleasure boat traffic on the river each summer is heavy. Note the lowered crossing gates on three of the four bridges, it looks like the gate on the left draw is stuck open. The massive coaling tower in the engine terminal is barely visible to the extreme left.

North Station was mostly eclipsed by the automobile, and as the early 50's passed, the new elevated expressway-cum-distressway rose in the background. The next photos show its tentacle-like growth. On May 24, 1952 the expressway bridge is rising over the open baggage door of B&M #6000 - *The First Streamliner East of the Mississippi!* Departure time is 4:00 PM for the MINUTE MAN on its daily 4 1/4 hour schedule to Troy, NY.

MARCH 21, 1936

Railway Age

Founded in 1856

What makes them Go-

EMC DIESEL POWER

Boston and Maine-Maine Central *"Flying Yankee"*—the first streamlined train on any Eastern railroad—celebrates its first birthday. Congratulations! "On Time" daily through winter blizzards as under summer sun—Over 250,000 miles of service—Boston-Portland-Bangor—726 miles each week day—over 95% availability. Such outstanding performance recovers passenger travel and increases revenues.

EMC

ELECTRO-MOTIVE CORPORATION
LA GRANGE, ILLINOIS, U.S.A.

Gas-electric car #187 burbles out of North Station toward Clinton, on the Central Mass Branch, October 9, 1954. A 1926 product of Osgood Bradley and Electro-Motive Corp. (the predecessor to EMD), the early Winton distillate engine produces 275HP. These Budd RDC predecessors last in Central Mass Branch service until 1956-or-1957, when RDC's replace them. The final coats of green paint are going on the expressway in the background.

On the further side of the draws, we find Pacific #3629 on the 1:40PM Lowell train. The reversed combine was common on B&M trains - smokers being segregated as far back as May 24, 1952! The bridge in the background carries the 3rd rail powered 'Subway' trains of the Charlestown Elevated out to Sullivan Square.

Pacific #3625 arrives at 8:50 AM with a Wilmington train. August 16, 1952 was a great day to visit North Station.

B&M P5 #3699 slowly backs down to its train at North Station on April 30, 1950. Originally built for the Lackawanna, it and 3 sisters arrived on the B&M in 1943 thanks to the crush of wartime traffic.

North Station meant switchers, and lots of them. Here #1108 deposits the CPR salon-observation assigned to the B&M/CPR RED WING service to Montreal. The imposing Massachusetts Registry of Motor Vehicles fills the background. The little SC arrived from EMC in 1938. Note ahead of the front walkway a gray New York Central wooden horse car on the further track. It likely came in from Troy on a Fitchburg Division train and is headed to Rockingham Park in Salem, NH. The date is June 6, 1954.

Alco RS-3 #1515 has just placed a draft of one or more 'American Flyer' passenger cars this June 19, 1955. If these were commuter or suburban coaches, likely #1515 would be the road power, but the main passenger trains still rate E- or F-units for a couple more years. The silver trucks are a nice touch. The concrete arches carry trolley cars to and from Cambridge.

Mistress of the Manor. E7 #3807 waits imperiously on October 11, 1947 for the conductor's whistle which will send her on the way to Portland or perhaps Montreal. Delivered in July 1946, she's recently been repainted from the grey backed 'Rock Island' scheme in which she was delivered.

Here we have a second view of E7 #3807 on April 30, 1950. In August 1949 - between views - #3807 had a head-on collision in Canaan, NH. with F2A #4225. It returned from La Grange with a new E8 cab sporting large, slanted numberboards. The wooden Hoods Milk car immediately behind the power suggests that this is the northbound ALLOUETTE, departing daily @ 9:00 AM for Montreal, PQ.

NW-2 #1208 is about to pull back into the yards with a wood-sided milk reefer on June 19, 1955. A Canadian Pacific smooth-sided baggage car is on the further track - likely about to leave for Montreal via Concord, NH; White River, Jct. and St. Johnsbury, VT. The REA packages are piled high on the carts this morning.

In a timeless scene, B&M Pacific #3606 slowly backs its commuter train of wooden cars from the passenger yard to North Station. Both cars and locomotive are pre-WWI era anachronisms on this October 11, 1947.

Nearly seven years later - October 9, 1954, Pacific #3627 is backing its train past Tower A and through the draws to its waiting passengers. These cars are steel, and the expressway menaces in the background, but still Steam Survives!

A month later, Nov. 6, 1954, #3639 chuffs slowly in reverse with its cars into North Station. The Shadows are long in many ways. The water tower so clearly visible in the background is on the roof of #150 Causeway Street - B&M headquarters next to North Station.

P2 #3625 backs its cars toward the passenger yard and out of North Station on a hazy August 16, 1952. Those high yard lights were excellent for fantastic photos - if you had the nerve to climb them!

B&M Pacific #3629 sits with its inbound train at North Station. The fireman's taking advantage of the lull to read the August 16th, 1952 edition of the paper.

A visitor is evident on the gangway of #3670. A conductor is likely giving him a disapproving glance this June 12th, 1952. Oh, that we could do it again.

Trackside at Boston Engine Terminal

North Station tracks cross the Charles River into Charlestown, and then split as they approach and pass the site of the Boston Engine Terminal, generally known as BET. Lawson shows us the BET in the waning days of Steam. The workmen were indulgent to fans, and the RRE arranged several tours of the facility during these postwar years. One of those trips was on October 2, 1954, when Lawson took many of these BET images. Most of the remainder were taken on August 30, 1947, likely an RRE visit as well.

BET's massive multi-track coaling tower. Several steam switchers cluster under the further side of the tower, with P5a Pacific #3699 in the foreground. Nary a diesel is visible this August 30, 1947.

With the yard as background, Pacific #3660 exits right (toward North Station) with spectacular smoke effects. Entering from the extreme right (toward the enginehouse) is SW-8 #803 on October 12, 1954.

One of several traded slides taken by fellow fan Leo Onofri. A nice near-broadside shot of B&M Atlantic #3235, now in 1948 reduced to commuter work. B&M #3235 was the road's last Atlantic, lasting into 1952.

B&M #3660 stands with the coaling tower and a GP-7 in the background on October 2, 1954. The tall brick stack in the background with the B&M so clearly visible upon it, marks the roundhouse powerhouse. Built in the late 1920's, the powerhouse supplied steam and electricity to BET - at 25 Hz, the industrial AC standard of the day. The 25 cycles-per-second (Hertz) flicker of the AC lighting was visible when walking into the house, but soon one's eyes compensated for it.

That same day Mogul #1388 stands beside a Pacific at the BET coaling tower. Let us remember these locomotives this way, alive, not being cut up at Billerica Shops.

With the New Hampshire Division 'High Line' as background, gas-electric car #1182 slumbered behind BET on August 30, 1947. The gasoline pump for it and its several remaining sisters was in this area, and is likely hidden behind #1182

Pacific #3625 has a full tender of coal for at least one more day of work before the scrap line.
Its bracketing between *Minute Man*-painted E7 and an Alco RS unit is accidental - but prophetic.

Mogul #1388 and an unidentifiable Pacific snooze by the coaling tower on October 2, 1954. The days are few for the Moguls, and the same goes for E7 #3811 getting a regular wash and polish job. Pat McGinnis with his hatchet and blue paint is in the wings.

A few minutes later we find the last operable B&M Consolidation - K7 #2403 - spotted at the coaling tower. The 2-8-0 finished its days on the Lexington Branch.

Turning from the coaling tower we see SW-8 #803 in the nearby yards. It was delivered by EMD in 1953. The large numberboards and MU are extra-cost options, yet its overall operating cost is about 1/3rd that of the 2 steam switchers it replaced.

Trackside up to the New Hampshire "North Country"

The B&M had an absolute transportation monopoly across New Hampshire. With miles of trackage, it dominated the state for a century. Many and frequent were the post-war Railroad Enthusiasts Trips crisscrossing the Granite State using otherwise idle commuter equipment.

Let's join Lawson for several visits 'Way up North - Way back When'.

Trip #1. April 30, 1950 - Boston to White River, Jct., via Concord and Claremont. No trip photos were taken between Boston and Concord, so we'll begin with a few individual scenes alongside the original Boston & Lowell/Lowell & Nashua/Concord Railroad trackage of circa 1840.

Late in Lawson's career on June 7, 1975, he visited the B&M's main shops in Billerica, 3.8 miles south of Lowell. Started about 1911, the shops were originally 'out in the middle of nowhere', and designed to be self-sufficient, with their own coal-fired powerplant for both electricity and heating. Coal fired production of 25-cycle power lasted until 1973.

Here's one of only five Budd RDC-4's (baggage-RPO only) ever built. Stored at Billerica shops for years, XKU 9081 was sold by the Minneapolis and St. Louis RR to Krautkramer Ultrasonics in 1967, who planned to make it and a sister into rail test or display cars. The project died, and the B&M acquired the car for non-payment of storage fees, and slowly stripped it for parts.

B&M's first E7, #3800, rolls west by timetable and north by compass past the North Chelmsford tower on August 8, 1948. Here the Stony Brook branch is about to carry the Washington-bound EAST WIND down to Ayer, where it will take the Worcester Branch to its New Haven connection in that city. Patronage is down nearly 30% from 1945.

The Boston-Montreal AMBASSADOR rolls north beside the Merrimack River between Lowell and North Chelmsford. E7 #3812 does the honors this sunny August 8, 1948.

Exactly 73.3 miles north of Boston is New Hampshire's State Capitol in Concord. Concord's large station - originally the offices of predecessor Concord RR - and its massive trainshed are the center of railroad activity. Northbound P5a Pacific #3699 pauses on the lead tracks just south of the station after bringing up our RRE special from Boston. It's a 'boomer' from the Lackawanna RR, arriving with sisters #3696-98 on the B&M in 1943. The Speed Lettering adds a nice touch of color this April 30, 1950. The P5a is too heavy for the branchline trackage west 57 miles to Claremont, NH, on the Connecticut River.

Concord to Portsmouth Train #3604 consists of EMC-Winton gas-electric #195 and trailer #180 on October 11, 1952. The train is passing through Hooksett, some 8 miles south of Concord.

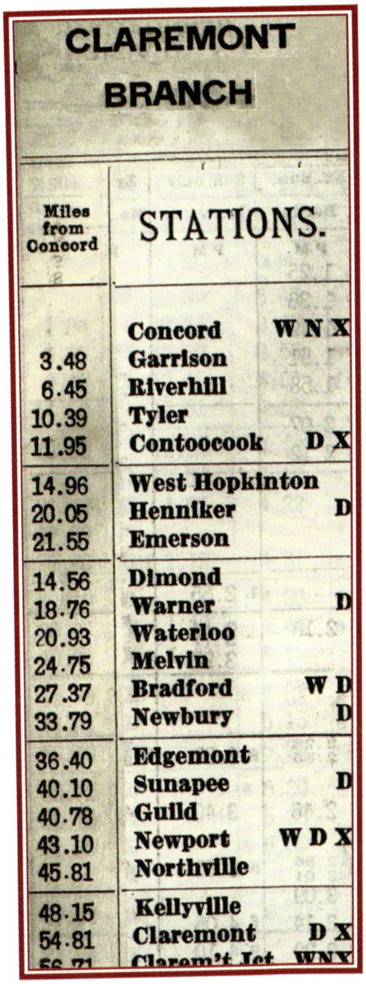

CLAREMONT BRANCH

Miles from Concord	STATIONS.	
	Concord	W N X
3.48	Garrison	
6.45	Riverhill	
10.39	Tyler	
11.95	Contoocook	D X
14.96	West Hopkinton	
20.05	Henniker	D
21.55	Emerson	
14.56	Dimond	
18.76	Warner	D
20.93	Waterloo	
24.75	Melvin	
27.37	Bradford	W D
33.79	Newbury	D
36.40	Edgemont	
40.10	Sunapee	D
40.78	Guild	
43.10	Newport	W D X
45.81	Northville	
48.15	Kellyville	
54.81	Claremont	D X
56.71	Clarem't Jct.	WNX

Quintessential B&M! The little town of Contoocook is 12 miles west of Concord, NH. Mogul #1465 heads the fantrip runby out of the venerable span. The well-maintained bridge still exists today, a civic centerpiece for this picturesque village. The trackage and Mogul aren't so fortunate.

Doubleheaded! Mogul #1482 assists #1465 with lots of smoke on the special through the Contoocook Covered Bridge.

Exiting the bridge, #1465 has its white flags (Running *Extra* - not scheduled service) in full sail. Looks like there are guests in the cab, and the fireman's taking in the fresh country air.

The doubleheader approaches the Kellyville Covered Bridge. Note that the bridge uses both arch and truss construction, with the arch footings visible on the stone piers beneath the structure itself.

It's now clear why a doubleheader is needed. What a multitude of aged commuter coaches and enthusiastic passengers.

Claremont Jct., where the branch runs alongside the tracks of the north-south Connecticut River division.

Lawrence is 26 miles north of Boston on the Western Route. Mogul #1488 gets its coal topped-off at the enginehouse before heading up the Lawrence branch to Manchester.

Trip #2. Boston to Concord (via Lawrence). Mogul #1488 is anxious to depart North Station with its contingent of fans this August 16, 1952. With headlight on, E7 #3820 is itself about to leave.

K7 Consolidation #2413 slumbers on the Lawrence ready track. A little more local and/or switching service awaits the aging steam power.

Slumbering ahead of Consolidation #2413 is Pacific #3676. Lawson is apparently standing on the lip of a gondola. These simple facilities were found across the B&M for basic servicing of its steam locomotive fleet.

New Hampshire's largest city is Manchester, home of the Amoskeag Mills, stretching some 7,500 feet along the east side of the Merrimack River. Little 0-6-0 switcher #441 rolls across the bricked-in turntable lead area in Manchester. The tall mill stack in the background speaks to the already passing age of steam and steel.

Southbound passenger Train #316 leaves Manchester for Boston behind GP-7 #1563. Note the several headend cars converted from WWII troop sleepers.

The road power for today's special, Mogul #1488 simmers contentedly, but no more than the fan in the fireman's seat! The 2-6-0 was delivered by Manchester in March, 1910, and has less than a year before scrapping.

Trip #3. June 14, 1953, a 'Railroad Enthusiasts' excursion to the North Country beckons. E7's #3807 & 3818 have the 'American Flyer' coaches well in hand. The entire consist is freshly washed and cleaned. The expressway ramps are growing like a fortress in the background, and the building on the right is the Registry of Motor Vehicles. (In November, 1999, these ramps had their 'Big Dig' successors partially open. Soon - like the Pacifics and Moguls below them in the 1950's - they too will fall to the scrapper.).

E7 #3807 and sister 3818 rest from their labors just north of White River Jct. on the CV line to St. Albans, VT. and Montreal. The 2000hp locomotives came from EMD in 1946 and 48 respectively.

The White River Jct. depot on June 14, 1953 still has lots of baggage carts and its famous locomotive weather vane. Unfortunately, the weather vane has been stolen several times over the years, most recently disappearing in the late '70's, apparently for good.

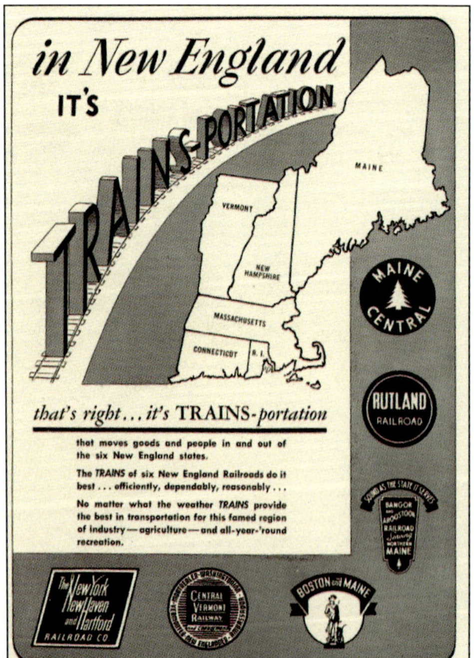

With a change of ends and run-around-the-train, the E7's approach the ball signal at White River. At night the red balls are replaced by lighting the kerosene lanterns mounted on the opposite side of the pole. The trackage facing us is the Connecticut River B&M trackage south from Wells River and the CP route from Montreal. The coaches behind the E's are still on Central Vermont trackage, and the train is about to cross the Connecticut River and head for Boston via Concord on the Northern line. Behind us the CV line continues down the Connecticut River Valley for 15 miles, where it becomes the B&M's Connecticut River line to Springfield.

The B&M's enginehouse was across the river in Westboro, NH. The E7's and 'American Flyer' coaches pause so all can get a look. The late afternoon sun illuminates the equipment beautifully.

Month old RDC-3 #6301 slumbers on the Westboro house tracks. It's the last reincarnation of the branch line local: RPO/Baggage/Coach in one carbody. In a decade, only North Station commuter service will remain.

With our visit to White River complete, it's time to examine the White Mountain Division to the north, then travel the Conn River line south, route of the fabled international night trains, the MONTREALER, and WASHINGTONIAN.

On July 21, 1953 one of B&M's four BL-2s poses beside the famous ball signal at Whitefield, NH. B&M #1550 ran until circa 1959; it was traded to EMD against a new GP-18 in 1961. The ball signal still stands in 2000!

At Thetford, Vt., some 12 miles north of White River, double-bladed 'Style B' lower quadrant semaphore signal #P133 indicates 'High Green' - maximum permissible speed.

P2 Pacific #3654 approaches with the southbound DAY WHITE MOUNTAINS EXPRESS and automatic block signal #P133 now stands 'Red'. It is August 6, 1949.

At Claremont Jct., 34.1 miles south of Thetford, this wonderful period image unfolds. A gas-electric - the connection east to Concord - and a bus await passengers at the station on July 23, 1953. Is the DAY WHITE MOUNTAIN EXPRESS late, perchance?

Minute Man clad F2 #4225 A&B, a southbound passenger job, is stopped at Claremont Jct. Possibly it is Train #72, the DAY WHITE MOUNTAIN EXPRESS, or #74, the CONNECTICUT YANKEE. This image is undated, a very unusual oversight for Lawson.

6Petite 44-tonner #110 switches the branch at Claremont Jct. on July 27, 1953. Soon this branch will become Sam Pinsly's Claremont & Concord RR.

At East Northfield, where the B&M Conn River line crosses the CV line to New London, CT., gas-electric car #195 awaits its Conn River connection. It soon will burble toward Keene, NH with (hopefully) several passengers.

The next station south of East Northfield is Mount Hermon; named for the Christian boy's school founded by Evangelist Dwight D. Moody in the late 1880's in Northfield. Mount Hermon remains today a top Prep School, though long merged with its feminine affiliate into Northfield Mount Hermon.

Southbound Train #74, the CONNECTICUT YANKEE speeds by Mount Hermon station non-stop behind P4a Pacific #3711 on April 17, 1948. This local stop has but a year or two to go before the automobile obviates its need.

The Last Hurrah. On July 27, 1957 a single about-to-depart Budd Highliner symbolizes the closing of the era of B&M passenger service to up-country New Hampshire. Boston service will just barely outlast the circa 1960 demolition of the Concord train shed and station. The traffic now passes by on four-lane I-93.

Trip #4. Snow Train! A Minute Man Budd RDC posed at the Franklin Jct. end of the Tilton-Frankin branch. The branch connects the Northern line from White River Jct. with the line up to Plymouth and Lincoln.

The branch to Lincoln, in northern New Hampshire, was a wintertime fantrip favorite. At the Tilton end of the Franklin-Tilton branch, excursionists disembark to stretch their legs on a wintry March 6, 1955.

Lincoln, where winter clearly lingers, and the paper mill's East Branch & Lincoln 0-4-0 tank engine is steamed up for the occasion. It appears completely work-a-day, but that won't last long.

The temporary Lincoln 'People Mover'! Have we not lost something with our obsession for liability suits??

EB&L #3 in a classic side shot. The mill - though without steam - lingered on into the 1970's before closing. Lincoln is now a tourist center, with Loon Mt. Ski Area just two miles from town.

Late in the afternoon of March 6, 1955, the Budd RDC's are about to depart from this scene of snow and steam. Note that two switchers have been fired up for the occasion.

In the grand tradition, Mogul #1421 backs down to its train on June 6, 1954. Powering an excursion; we shall soon see it again.

Trackside on the Central Mass Branch

THE CENTRAL MASS BRANCH WAS A LINE THAT NEVER SHOULD HAVE BEEN BUILT. CONCEIVED IN THE post-Civil War era, it paralleled the B&M/Fitchburg line to its north, and the B&A line to its south, eventually reaching the Connecticut River Valley at Northampton. It slowly withered from the center to both west and east - the Hurricane of '38' helped greatly. By 1951 Clinton - 37 miles from Boston - was served by only four daily round trips.

In the early '50's the Central Mass Branch was a time machine. Equipment was mid-20's vintage EMC/Winton/Osgood Bradley gas-electrics, or 'Teens' Moguls with wooden open vestibule coaches. B&M Moguls held their last stand here on May 5, 1956. Legend says that as commuter train #3106 passed over the North Station draw, its fireman tossed his shovel into the water to mark the occasion! The gas-electrics soon disappeared as well, both being replaced with Budd RDC's. Service to Clinton ended in 1958, but continued on to South Sudbury until November 26, 1971. Some of Lawson's best work was done here.

Four car Train #3111, the 5:17 PM, rolls west of Waverly behind 2-6-0 #1470 on July 24, 1951.

CENTRAL MASSACHUSETTS BRANCH.

	Miles from Boston	STATIONS.		
TERMINAL DIVISION		**Boston**		W N X
	4.16	**West Cambridge**		T
	4.57	**Fens**		T
	5.52	**Hill Crossing**		T
	6.42	**Belmont**		
	7.38	**Waverley**		T
	8.30	**Clematis Brook**		T
	9.77	**Waltham (North)**		W
	10.35	**Waltham Highlands**		
	12.93	**Weston**		D
	13.69	**Cherry Brook**		
	15.24	**Tower Hill**		
	16.50	**Wayland**		D
	18.55	**East Sudbury**		
	19.70	**South Sudbury**		D
	21.63	**Wayside Inn**		
	23.84	**Ordway**		
	25.35	**Gleasondale**		
	26.09	**Gleason Junction**		
		C. M. Jct.		
	27.69	**Hudson**		W
	29.49	**South Bolton**		
	31.42	**Berlin**		
	35.34	**Reservoir Switch**		T
	36.00	**East Switch**		T
	36.50	**Clinton Junction**		T
FITCHBURG DIVISION	65.45	**Wheelwright**		
	68.33	*Creamery*	B.&A.R.R. #	
	70.58	*Gilbertville*	B.&A.R.R.	
	74.59	*Ware*	B.&A.R.R.	W
	79.07	*Forest Lake*	B.&A.R.R.	
	82.27	**Bondsville**		
	85.87	*Canal Jct.*	C.V Ry.	
		Belchertown	C.V Ry. #	
	94.17	*Norwottuck*	C.V Ry.	X
	97.19	**Amherst**		D
	101.50	**Hadley**		
	103.77	**N. O. Tower**		T W D X

On August 20, 1952, Train #3117 passes through the new crossover at Clematis Brook behind Mogul #1495, one of the last two 2-6-0's in service. The Central Mass trackage that parallels the Fitchburg Division east for 3.73 miles to Fens is now abandoned.

With shadows lengthening, a few minutes later the three cars of Train #3117, the 5:50 PM, pass Clematis Brook behind 2-6-0 #1388.

That same afternoon Mogul #1425, bound for Maynard via the Fitchburg Division to South Acton, roars up the main with five car Train #5205.

Three days earlier, July 21, 1951, the four cars making up Train #3109 pass a scenic spot in Waltham behind the last Atlantic: #3235. This is the only remaining service to Lancaster, past Clinton Junction north 2.4 miles up the Worcester main toward Ayer.

An absolutely gorgeous shot of Train #3111 and Mogul #1420 taken but a few minutes after the passage of Train #3109.

At the height of fall foliage on October 9, 1948, Atlantic #3235 and three-car Train #3105 to Lancaster approach a crossing in Waltham.

Train #3105, also in Waltham, rolls through a stark field on November 20, 1948. Atlantic #3232 heads the 3-car train in this bucolic scene.

On June 6, 1954 #1431 slumbers on the Clinton enginehouse ready track. Likely it will leave @ 6:18AM Monday morning as Train #3100 filled with Boston-bound passengers.

Mogul #1421 pauses on the stone arch bridge over the spillway at Wachusetts Reservoir on June 6, 1954. The trackage has been gone since 1958, but the stone arch bridge remains.

Thanks to the construction of the massive Clinton dam and Wachusetts reservoir between 1896 and 1903, the Central Mass acquired this massive viaduct as part of its track relocation early in the 20th Century. Mogul #1421 and its train of excursionist-filled 'American Flyer' coaches ride high some 300 yards or so in front of the stone dam, fountain and gardens.

The *First Streamliner East of the Mississippi* rolls past Tower A beginning its 4 hour and 15 minute trek to Troy, NY as the MINUTE MAN on November 6, 1954.

Trackside on the Fitchburg Division

THE FITCHBURG DIVISION WAS THE B&M'S MAIN LINE west to the Hudson River Valley at Mechanic-ville, NY - north of Albany. The end of Boston suburban service was Fitchburg, 49.6 miles west.

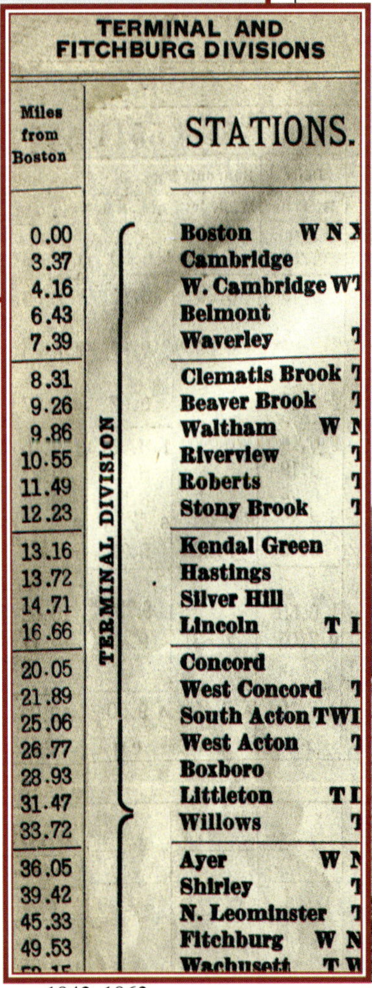

TERMINAL AND FITCHBURG DIVISIONS

Miles from Boston		STATIONS.
0.00	TERMINAL DIVISION	Boston W N X
3.37		Cambridge
4.16		W. Cambridge W T
6.43		Belmont
7.39		Waverley T
8.31		Clematis Brook T
9.26		Beaver Brook T
9.86		Waltham W N
10.55		Riverview T
11.49		Roberts T
12.23		Stony Brook T
13.16		Kendal Green
13.72		Hastings
14.71		Silver Hill
16.66		Lincoln T I
20.05		Concord
21.89		West Concord T
25.06		South Acton T W I
26.77		West Acton T
28.93		Boxboro
31.47		Littleton T I
33.72		Willows T
36.05		Ayer W N
39.42		Shirley T
45.33		N. Leominster T
49.53		Fitchburg W N
[illegible]		Wachusett T W

The previous September 20th a local freight passes by the work at Waverly behind 0-6-0 switcher #431. It's on the soon-to-be abandoned Central Mass branch alignment

On February 14, 1953, at the height of the work to lower the Fitchburg main through the center of Waverly, B&M #6000 rolls west as the MINUTE MAN.

The 5:14 to Fitchburg is west of Waverly in the capable hands of a 3700-class P3 Pacific on July 24, 1951. The track in the foreground is the Central Mass branch.

Train #511, a westbound Fitchburg local, rolls through the snowy landscape of Waltham on February 21, 1948. The three rear coaches are ex-PRR cars recently purchased by the B&M.

At Clematis Brook Train #5507 heads to Bellows Falls, via Keene, on the Cheshire Branch. It's October 9, 1948.

Another view of Train #5507 at Clematis Brook, this time on a snowless December 30, 1951. From the amount of smoke, it's clear the fireman is pouring on the coal.

On February 21, 1948, Lawson found Train #509 behind Mogul #1498 in a sunnier spot. This and the prior photo were taken near the present location of the Polaroid building and Route #117 crossing of Route128/I-95. How quickly this rural landscape faded away.

On a sunny and beautiful July 21, 1950, Fitchburg local #505 rolls through the outskirts of Waltham behind P2 Pacific #3683.

After a few minute's wait, the scene is repeated with Train #515. On the left are the tracks of the Central Mass branch.

Another short hiatus, and Train #517 with #1425 rolls into view at the height of the commuter rush hour.

The Maynard branch local, Train #5205, marches through Clematis Brook behind Mogul #1495. Scheduled arrival is 6:pm.

The returning local freight rolls toward Boston behind Consolidation #2688. The outside braced wooden B&M boxcar is a study all its own.

Waltham station, a location Lawson knew well. On March 20, 1948 Pacific #3680 awaits the conductor's whistle with Train #5507 to Rutland, VT. via the Rutland RR from Bellows Falls, VT.

Proof that the B&M power worked through to Rutland, VT: P3 Pacific #3704 prepares to depart the Rutland yard for its train on September 21, 1951.

Hurricanes devastated the B&A through the Berkshires in both 1954 and 1955. The detoured B&A NEW ENGLAND STATES rolls through Waltham on September 11, 1955. Twin E7A's led by #4022 head the westbound streamliner.

East of Roberts (1.6 miles west of Waltham) on the same day, B&A Train #33, the Chicago-bound NEW ENGLAND WOLVERINE rolls west over the B&M behind matched E8A's.

A westbound for Troy stands at Waltham station on Independence Day, 1957. Gone are steam, gas-electrics, and even the Minute Man paint - replaced by the dubious esthetics of the new McGinnis 'Black and Blue' scheme. Troy service has little more than six months to go. B&M #3814 is the only E7 to wear this scheme.

0 Boston to Troy, N. Y.

For complete service Boston to F

				WEEKDAYS							SUNDAYS		
iles				◆523	51	◆55	57	◆59	◆61	◆527	67	◆69	◆521
				A M	A M	A M	A M	P M	P M	P M	A M	P M	P M
.0	BOSTON (No. Sta.)	Mass.	Lv	12 15	2 00	8 15	10 00	5 00	‡6 05	k11 55	8 30	5 55	9 30
.4	Cambridge	"	......	......	......	8 23	10 08	v5 07	‡6 13	▲12 02	8 38	6 02	9 37
.9	Waltham	"	...	...	2 18	8 33	10 20	a	‡6 21	▲12 11	8 50	6 10	9 46
.1	Concord	"	......	......	2 31	8 49	10 38	......	‡6 36	▲12 25	9 05	6 22	10 00
.1	Ayer	"	...	...	3 03	9 07	11 15	5 43	‡7 02	▲12 48	9 32	6 40	10 20
.6	Fitchburg	"	Ar	......	3 27	9 28	11 41	6 03	‡7 24	▲1 08	9 54	6 55	10 37
.6	FITCHBURG	"	Lv	v1 12	3 35	9 32	11 51	6 06	‡7 24	▲1 12	10 00	6 55	10 37
.9	South Ashburnham ◇	"	......	......	......	......	12 10	......	‡7 e41	......	......	......	......
.1	East Gardner ◆	"	...	...	...	...	f12 16	...	...	...	...	...	...
.7	Gardner	"	Ar	......	4 00	9 55	12 20	6 29	‡7 49	......	10 24	7 17	10 58
.7	GARDNER	"	Lv	f1 36	4 15	9 55	12 42	6 29	‡7 49	▲1 f36	10 34	7 17	10 58
.5	Baldwinville ◇	"	......	......	......	f10 02	12 52	......	‡7 e56	......	......	......	......
.4	Royalston ◆	"	...	...	...	...	1 00	...	‡8 f02	...	...	...	...
.7	Athol	"	......	e1 58	4 44	10 20	1 25	6 52	‡8 13	▲1 e58	10 59	7 41	11 20
.0	Orange	"	...	g2 04	4 54	10 26	1 35	6 58	‡8 19	▲2 g04	11 05	7 47	e11 26
.0	Erving	"	...	...	...	...	1 45	...	...	...	...	...	...
.1	Millers Falls	"	......	......	5 11	f10 40	1 55	......	‡8 e34	......	......	......	......
.9	Lake Pleasant ◆	"	...	...	5 15	...	f1 59	...	...	...	...	...	...
.5	Montague ◆	"	......	......	5 19	......	2 05	......	......	......	......	......	......
.4	Greenfield	"	Ar	2 30	5 29	10 55	2 15	7 25	‡8 48	▲2 30	11 31	8 11	11 52
.4	GREENFIELD	"	Lv	2 30	6 00	11 20	3 00	7 30	...	▲2 30	11 45	8 25	...
.4	Shelburne Falls ◆	"	......	......	6 19	f11 38	3 20	7 48	......	......	12 03	8 43	......
.1	Charlemont ◆	"	...	...	6 32	f11 50	3 35	e7 59	...	...	...	...	...
.6	Zoar ◆	"	......	......	e6 38	......	......	......	......	......	......	......	......
.7	Hoosac Tunnel ◆	"	...	...	6 45	f12 04	3 50	...	...	...	...	...	...
.2	North Adams	"	Ar	......	7 02	12 20	4 10	8 26	......	......	12 42	9 22	......
.2	NORTH ADAMS	"	Lv	f3 27	7 10	12 20	4 30	8 26	...	▲3 f27	12 50	9 22	...
.8	Williamstown	"	......	e3 34	7 20	12 27	4 42	8 33	......	▲3 e34	12 58	9 30	......
.8	North Pownal ◆	Vt.	...	...	7 31	...	4 52	...	...	...	...	...	...
.0	Hoosick ◇	N. Y.	......	......	7 45	......	......	......	......	......	......	......	......
.2	Hoosick Falls	"	...	...	8 05	12 47	5 20	8 53	...	...	1 20	9 50	...
.3	Eagle Bridge	"	......	......	8 12	f12 53	5 41	......	......	......	......	......	......
.3	⊙East Buskirk ◆	"	...	...	......	...	f5 45	...	...	...	...	...	...
.9	Johnsonville ◆	"	......	......	8 23	f1 01	5 53	e9 06	......	......	f1 33	......	......
.6	Valley Falls	"	...	...	8 28	f1 05	5 58	...	...	...	...	...	...
.2	TROY	"	Ar	......	8 55	1 25	6 20	9 30	......	......	1 59	10 25	......
	TROY	"	Lv	...	♀	♀	♀	♀	...	...	♀	♀	...
.5	ALBANY, NYC RR.	"	Ar	......	♀	♀	♀	♀	......	......	♀	♀	......
				AM	AM	PM	PM	PM	PM	AM	PM	PM	PM

a Stops to receive passengers 5 15 p.m.

e Stops to discharge passengers

f Stops on signal to discharge o

g Stops to discharge passenge bridge.

h Flag stop Friday only at Zoa

k Saturday only.

v Stops to receive passengers.

‡ Except Saturday.

▲ Sundays only.

⊙ East Buskirk and Buskirk are about one-half mile apart; take trains at East Buskirk,

◆ No checked baggage handled referenced. Checked bagga trains referenced at top number.

◇ Tickets are sold, but no checked baggage handled at stations showing reference ◇.

♀ Buses of the United Traction Co. run at frequent intervals between Troy and Albany.

A Boston-bound freight passes Riverview, along the Charles River, .69 of a mile west of the Waltham station on February 21, 1948. Power consists of three F2's: A-B set #4225, and trailing A #4261. This lashup is unusual, as #4225 A-B (and its two sister sets) were purchased as upcountry passenger power, while 15 F2A's including #4261 were bought to break up the 12 FT ABBA sets. It was very common to see an FT AB/F2A set at this time.

With the image of a bright Minute Man-painted F-cab set in our minds, now is the appropriate time to leave the B&M and go across town to the B&A.

THE BOSTON & ALBANY MAIN LINE STRETCHES 200 miles between the bumpers of South Station and the Albany, NY Depot on the bank of the Hudson River. Of that total, slightly more than the 161.8 miles are from Boston to the Massachusetts border, just west of the flag stop of State Line, MA. Unlike the B&M and New Haven, the B&A main was not replete with branches or connections every few miles. In fact, there were but a handful.

The four-track main departing South Station was dispatched and operated as two double-track main lines. The southern main handled local service, and the northern one through service. The twin stations of Huntington (eastbound) and Trinity Place (westbound) were located only 1.28 miles west of South Station. All passenger equipment was stored and maintained in the Exeter Street yard, bounded by Huntington Avenue, and Exeter and Boylston streets.

The Grand Junction Branch left the main at the east end of the Beacon Park engine terminal, reached across the Charles River, through Somerville, across the B&M yards behind Boston Engine Terminal, and circled around through Everett and Chelsea to reach the docks of East Boston - across Boston harbor. Between coastal shipping and the large number of industrial sidings on the Grand Junction, this undoubtedly was the B&A's most profitable branch.

Beacon Park was well equipped for engine service with two roundhouses and a massive coaling tower. Adjacent to it was the Allston freight yard. The B&A's main Boston freight yard, it contained 55 tracks with room for 2500 freight cars. This area stretched along the main between mileposts 3.32 and 4.24 beside the Charles River.

About a mile west of Beacon Park, at Brookline Junction the Newton Highland Branch to Riverside left the main. This branch handled significant B&A commuter traffic from Brookline and Newton until its conversion to trolley service by the MTA (Metropolitan Transit Authority) in 1958-59.

On the east side of Worcester, at Millbury Jct., a branch connected with Millbury, MA. About five miles west of Worcester, at Webster Jct, the Webster Branch headed for Webster, MA., and another New Haven connection.

Further west at South Spencer, milepost 69.1, was the branch to Spencer, and at East Brookfield, milepost 69.6, was the branch to North Brookfield.

A substantial B&A branch stretched northwest from Palmer, milepost 86.3, to Winchendon, MA. and a connection with the B&M. A similar line, the Athol branch, stretched from Athol Jct., just east of Springfield, up the Swift River Valley to the B&M's Fitchburg Division main in Athol. This line was abandoned beyond Bondsville in 1935 thanks to the construction of the massive Quabbin Reservoir to serve as Boston's main water supply.

Deep in the Berkshires at Pittsfield Jct., near milepost 149, the North Adams Branch headed for that city and another B&M connection.

Over in New York State the B&A had two branches, one from Chatham down to Hudson, on the New York City main, and the freight branch - now the main - from Post Road over the massive Al Smith Bridge above the Hudson River to Selkirk Yard. This was, and is, the confluence of the NYC (now CSX) West Shore line, the B&A main, and the Water Level (freight) route.

B&A Steam Locomotive Roster Circa 1941-1951 (End of steam)

TYPE	CLASS	NUMBER	DATE(S) BUILT	BUILDER
0-6-0	B10e/i/m/o/s	#115-143	1907-1913	Alco: Sch/RI/Manch
	B11l/o	#144-152	1913/1916	Alco: Schenectady
0-8-0	U33	#30-41	1901-02	Alco: Schenectady
	U2j/3b/2k/2l	#42-65	1917-1924	Alco: Schen/Lima
2-6-6T	D2a	#300-309	1906/rebuilt 1928-31	Alco: Schen/Lima
	D2b	#310-317	1906/rebuilt 1928-31	Alco: Schen/Lima
4-6-6T	D4	#400-404	1928	Alco: Schenectady
4-6-2	K3n	#500-506	1918	Alco: Brooks
	Kj/l/m	#550-565	1911-1914	Alco: Schenectady
	K14g(exNYC)	#575-588	1913	Alco: Schenectady
	K14h(exNYC)	#589	1911	BLW
4-6-4	J2a	#600-604	1928	Alco: Schenectady
	J2b	#605-609	1930	Alco: Schenectady
	J2c	#610-619	1931	Lima
2-8-0	G6h	#1030-1049	1908	Alco: Schenectady
	G6q	#1050-1053	1912	Alco: Schenectady
2-8-2	H5j	#1200-1213	1913-1914	Alco: Schenectady
	H5g	#1214-1223	1915	Alco: Brooks
	H5t	#1224-1229	1916	Lima
2-8-4	A-1a	#1400-1424	1926	Lima
	A-1b	#1425-1444	1926-1927	Lima
	A-1c	#1445-1454	1930	Lima

New York Central Steam Locomotives frequently used on the B&A

TYPE	CLASS	NUMBER	DATE(S) BUILT	BUILDER
4-6-2	K11a	#4486/4498	1911-1912	Alco: Brooks
	K3p	#4697-4722	1920	Alco: Schenectady
	K3r	#4800-4804	1911	Alco; Brooks
4-6-4	J1a	#5200	1927	Alco: Schenectady
	J1b	#5201-5249	1927	Alco: Schenectady
	J1c	#5250-5274	1928-1929	Alco: Schenectady
	J1d	#5275-5314	1929-1930	Alco: Schenectady
	J1e	#5315-5344	1931	Alco: Schenectady
	J3a	#5405-5444	1937	Alco: Schenectady
	J3a	#5445-5454	1937 (Streamlined)	Alco: Schenectady
4-8-2	L3a	#3000-3024	1940	Alco: Schenectady
	L3b	#3025-3034	1940-1941	Alco: Schenectady
	L3b	#3035-3049	1940-1941	Lima
	L4a	#3100-3124	1942-1943	Lima
	L4b	#3125-3149	1943-1944	Lima

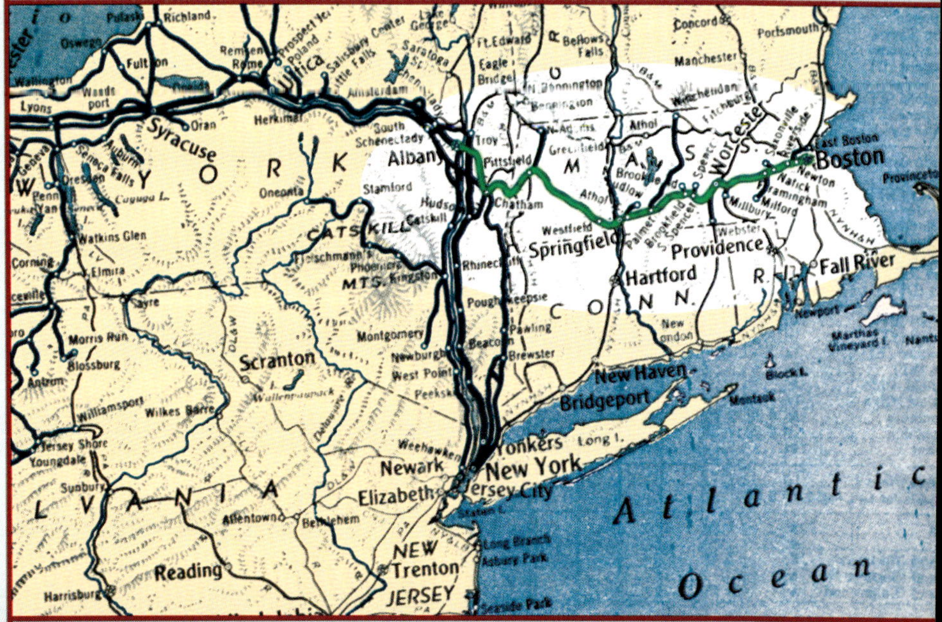

B&A/NYC Diesel-Electric Locomotive Roster circa 1956

Alco FA/FB, PA/PB, RS-2/-3, EMD E7-E8, and L-H RS units were the most common B&A power between 1945 and 1955. By 1960 the ALCO PA/B and L-H RS units were gone from the B&A.

NUMBER	BUILDER/TYPE	YEAR BUILT
#526-562	Alco/GE/I-R 'Three-Power'	1930
#567-573	EMC SC	1936
#574-704	EMC-EMD/SW-1	1939-50
#750-761	BLW VO-660	1941-44
#800-810	Alco HH-600	1938-39
#811-873 & #950-57	Alco S1	1941-50
#874-916	Alco S3	1950-51
#1000-43	Alco FA1	1947-49
#1044-1123	Alco FA2	1951-52
#1600-03	EMD FTA	1944
#1604-05	EMD F2A	1946
#1606-35	EMD F3A	1947
#1636-1873	EMD F7A	1949-52
#2300-22 (Renumbered to #3300-22)	Alco FB1	1947-49
#2400-03	EMD FTB	1944
#2404-19	EMD F3B	1947-48
#2420-74	EMD F7B	1949-52
#3200-03	BLW DR6-4-1500 (A-1-A)(Passenger A-cab)	1947-48
#3300-01 (renumbered to #3400-01)	BLW DR6-4-1500 (A-1-A)(Passenger B unit)	1947-48
#3323-72	Alco FB2	1951-52
#3400-03 (renumbered to #3800-03)	BLW DR 4-4-1500 (Babyface freight A-cab)	1948
#3500-03	EMD F3A (Passenger)	1947
#3600-01	EMD F3B (Passenger)	1947
#3700-01	BLW DR 4-4-1500 (Babyface freight B unit)	1948
#3702-09	BLH RF-16 (Sharknose B unit)	1951-52
#3804-21	BLH RF-16 (Sharknose A-cab)	1951-52
#4000-35	EMD E7A	1945-49
#4036-95	EMD E8A	1951-53
#4100-13	EMD E7B	1945-48
#4200-07	Alco PA-1	1948-49
#4208-14	Alco PA-2	1950-52
#4300-03	Alco PB-1	1948
#4304-04	Alco PB-2	1950
#4400-05	FM Erie-built (Passenger A cab)	1949
#4500-7	FM CPA 24-5 (C-liner Passenger A cab)	1952
#5000-05	FM Erie-built (Freight A cab)	1947-49
#5006-17	FM CFA 20-4 (C-liner Freight A cab)	1950
#5100-01	FM Erie-built (Freight B unit)	1948-49
#5102-04	FM CFB 20-4 (C-liner Freight B unit)	1950
#5600-5827	EMD GP-7	1950-53
#5900-6028	EMD GP-9	1955-56
#5800-14	L-H RS	1950
#5820-36	BLW RS-12	1951
#6900-03	FM CFB 16-4	1952
#7100-18	FM H20-44	1948-49
#8200-22	Alco RS-2	1948-50
#8223-8357	Alco RS-3	1950-53
#8400-11	L-H 1000HP switcher	1949-51
#8500-89	Alco S2	1943-50
#8590-8667	Alco S4	1952-53
#8600-07 (renumbered to #9300-07)	BLW VO-1000	1944-45
#8700-8834	EMD NW-2	1946-49
#8835-8921	EMD SW-7	1950-51
#8922-9008	EMD SW-9	1951-53
#9100-10	FM H10-44	1946-50
#9111-37	FM H12-44	1950-52
#9500-16	EMD NW-2	1948
#9800-20	L-H 800HP switcher	1951

Trackside on the main from Boston to Riverside

OUR B&A VISIT BEGINS ON THE GRAND JUNCTION BRANCH IN EAST BOSTON.

The East Boston B&A grain elevator as seen from Boston Harbor on August 8, 1964. Millions of tons of grain passed from the Midwest via the NYC/B&A through this elevator to ocean-going vessels. South Station was magnificent, but tonnage paid the bills.

Tonnage from East Boston came via the Grand Junction, but people crossed the harbor on the East Boston ferries. Although not a B&A-operated affair, still they were a unique part of the Boston harbor scene for over a century. Here ferry *Daniel A. MacCormack* is docked at East Boston.

Two views of sister ferry *Charles C. Donoghue* as it exits the Boston pier and heads across the channel. The B&A grain elevator is clearly visible in the distance. After being put out of work in Boston, the two provided ferry service across the Hudson River between Beacon and Newburgh NY, until the I-84 bridges were opened.

May, 1964 means times are 'A-Changin' in Boston. Trinity Place station - named for the magnificent Trinity Church designed by H.H. Richardson in Copley Square - ponders its fate. Built by Richardson's successor firm shortly after his death in the 1880's, the station will shortly fall to the approaching Massachusetts Turnpike Extension. What little passenger traffic remains will be rerouted through the New Haven's adjacent Back Bay station.

Another view of Trinity Place and Back Bay stations, some 1.28 miles from South Station. The B&A's eastbound station - Huntington Avenue - has already been demolished, and replaced by the very evident Turnpike bridge pier construction. To our rear, the Exeter Street Coach Yard is metamorphosing into the multi-million dollar Prudential Center.

Cottage Farms, Brookline, was where the Grand Junction Branch connected with the B&A main - the east end of Beacon Park yard in Allston. Here the first section of #27 - the NEW ENGLAND STATES - rolls westbound behind #600, the B&A's first J2a Hudson. The wartime date is September 17, 1942.

Lawson visited Beacon Park on a bright September 17, 1949 and recorded the next four images. On the turntable, suburban 2-6-6T tank engine #312 takes a swing. The engine was built by Alco in 1906, rebuilt by Lima in 1928, and soon will be scrapped.

B&A 0-6-0 #140 also basks in the bright sun at Beacon Park. She too has little time left.

Consolidation #1052 has steam in its cylinders this early fall day. Note the old style low pressure 'city' or illuminating gas storage tanks in the background.

Fairbanks-Morse 2000 HP 'Erie-built' #5002 and B-unit #5100 or -01 show the road grime acquired doing battle with the hills of Massachusetts. B&A #5002-05A and #5100-01B, arrived in early spring, 1949. Note that the side stripe ends part way across, possibly to avoid covering the radiator shutters. Along with most of the 'Erie-built' fleet, NYC #5002 was shipped to EMD in the '50's for repowering with V-16 567 engines.

On September 12, 1948, two 4-6-6T suburban tank engines roll past Beacon Park tower.
Lawson is standing on the Allston depot platform. Note the flange oiler under the closer engine's drivers.

Corrected to July 21, 1918

UNITED STATES RAILROAD ADMINISTRATION

W. G. McADOO, DIRECTOR GENERAL OF RAILROADS

BOSTON and ALBANY RAILROAD

NEW YORK CENTRAL LINES

Time Tables

Boston, Worcester and Intermediate Stations

Milford and Webster Branches

E

STATIONS	3	663	63	21	65	7	9	67	71	51	41
	Dly. Ex. Sun.	Dly. Ex. Sun.	Dly. Ex. Sun.	Dly. Ex. Sun.	Dly. Ex. Sun.	Dly. Ex. Sun.	Sun. Only	Dly. Ex. Sun.	Sun. Only	Dly. Ex. Sun.	Dly.
			Note								
	AM	AM	AM	AM	AM	AM	AM	AM	AM	AM	AM
Boston........Lv.	4 05	....	6 30	7 10	7 30	8 30	8 30	8 50	9 15	9 15	10 00
Trinity Place......	4 09	....	6 34	7 14	7 34	8 34	8 34	8 54	9 19	9 19	10 04
Cottage Farm.....	....	....	....	....	....	....	....	....	....	....	....
Allston..........	4 15	....	6 40	....	7 40	....	....	9 01	9 26	....	....
Brighton..........	....	....	6 43	....	7 42	....	....	9 03	9 28	....	....
Faneuil..........	....	....	6 46	....	7 45	....	....	9 05	9 31	....	....
Newton..........	4 21	....	6 49	....	7 49	....	....	9 08	9 34	....	....
Newtonville......	....	....	6 52	....	7 53	....	8 45	9 11	9 37	x9 30	10 15
West Newton.....	4 26	....	6 55	....	7 57	....	....	9 14	9 40	....	....
Auburndale.......	....	....	6 58	....	8 00	....	....	9 17	9 43	....	....
Riverside........	4 32	....	7 01	....	8 03	....	....	9 20	9 46	....	....
Wellesley Farms...	....	....	7 05	....	8 07	....	....	9 24	9 50	....	....
Wellesley Hills....	....	....	7 08	....	8 10	....	....	9 27	9 53	....	....
Wellesley.........	....	....	7 14	....	8 13	....	....	9 30	9 57	....	....
Natick...........	4 46	....	7 22	....	8 19	....	....	9 37	10 03	....	....
Saxonville (Br.). Ar.	....	....	....	....	....	....	....	....	....	....	....
Framingham...Ar.	4 54	....	7 28	7 43	8 28	9 03	9 08	9 53	10 10	....	10 38
Framingham...Lv. (Branch)	6 05	....	7 50	7 50	....	....	....	....	....	....	....
Milton Br. Whitneys.........	....	....	7 54	7 54	....	....	....	....	....	....	....
Milton Br. E. Holliston....	....	....	7 57	7 57	....	....	....	....	....	....	....
Milton Br. Holliston.......	6 22	....	8 02	8 02	....	....	....	....	....	....	....
Milton Br. Metcalfs.......	....	....	8 07	8 07	....	....	....	....	....	....	....
Milton Br. Braggville......	....	....	8 12	8 12	....	....	....	....	....	....	....
Milton Br. Milford.....Ar.	6 35	....	8 20	8 20	....	....	....	....	....	....	....
Framingham...Lv. (Main Line)	4 54	5 35	....	7 43	....	9 03	9 08	....	10 10		10 38
Ashland.......Lv.	5 00	5 40	....	7 49	....	Buffet Club Car and Parlor Car.	....	....	10 16	Parlor Car and Coaches	Parlor Car and Coaches
Cordaville.........	....	5 46	....	7 55	....		....	....	10 21		
Southville.........	....	5 49	....	7 58	....		....	....	10 24		
Westboro.........	5 14	5 56	....	8 05	....		....	....	10 31		
North Grafton....	5 23	6 05	....	8 14	....		....	....	10 42		
Millbury Junc.....	....	....	....	8 17	....		....	....	....		
Millbury (Br.) .Ar.	....	....	....	8 45	....		....	....	....		
Worcester......Ar.	5 36	6 18	....	8 27	....	9 35	9 40	....	10 53	10 20	11 1
	AM	AM	AM	AM	AM	AM	AM	AM	AM	AM	AM

B Does not carry local baggage.
d Will not run between Framingham and Milford on Feb. 22, April 19, May 28, or Dec 25, 1918.
f Will stop on signal to receive or discharge passengers.
g Stops Monday mornings only.
k Runs Saturdays only.

Later that same day 'Sports model' A1c class 2-8-4 #1451 drifts by the depot. It still stands today, for over two decades preserved as first a restaurant, 'The Allston Depot', and more recently as the 'Allston Sports Bar'. Note the two dwarf semaphore signals.

B&A Train #31, a 'Saturday Only' train to Springfield, rolls through Brighton with NYC Pacific #4801 leading three coaches and a baggage car some two hours behind the ADVANCED KNICKERBOCKER.

The Boston Section of the ADVANCED KNICKERBOCKER rolls through Brighton at approximately 11:15 AM on Saturday, September 17, 1949. L3a Mohawk #3014 is in charge.

NYC L3a Mohawk #3005 is getting a roll on with Train #11 - the Boston Section of the St. Louis-bound SOUTHWESTERN LIMITED - through Brighton on July 2, 1949.

Riverside, where the B&A's Newton Highland commuter line and the main connect. On April 8, 1950, L3a Mohawk #3005 shows she can still do her assigned task with the westbound Boston Section of the KNICKERBOCKER. What a grand image of big time railroading with steam, steel, and semaphore signals.

On April 22, 1950, Lawson was able to study B&A Class D1a 4-6-6T tank engines #402 and #404 at Riverside. Here's a broadside of #402, likely while awaiting its trip to the scrapper.

With the solid frame carrying through under the tender, #404 clearly shows the miniature pilot and headlight necessary for reverse running at speed with regular 'loop' trains from South Station up the branch to Riverside, and then back via the main, or vice-versa. They were as bi-directional as a GP-7 or RS 2.

Four months after the diesels completed their conquest of the B&A, on August 19, 1951 the NEW ENGLAND STATES rolls west past Riverside behind E8 #4037 and a sister. The 4-track main and train still had panache.

Trackside on the Newton Highland Branch to Riverside

The Newton Highland Branch began in 1848 as the Brookline Branch of B&A predecessor Boston & Worcester. Extended to Needham by the early 1850's, much of the dirt used to landfill Boston's mosquito-laden Back Bay into plots for posh Brownstone residences traveled over these tracks during the next 25 years.

In final railroad form the Newton Highland Branch left the B&A main at Brookline Junction, about 2 miles from South Station, and meandered past numerous suburban stops in Brookline and Newton. Reaching Riverside, the branch rejoined the main on a 'U', making suburban or local 'circuit' trains possible. They began running in May, 1886, out one way, and back the other.

In 1906 the first 2-6-6T tank engines were delivered for the 'circuit' trains. With the tender bunker and tank mounted on an extension of the locomotive frame - and carried on a six-wheel truck - the locomotives were equally capable of forward or backward running at road speed. Five larger 4-6-6T's came in 1928. Tank engines handled the 'circuit' service through 1950, when Lima-Hamilton and Alco road switchers replaced them.

In May, 1958 the MTA (Metropolitan Transit Authority) purchased the Newton Highland Branch from the B&A and over the next 13 months converted it to trolley use. A short ramp and tunnel section was built to connect the line into the Green Line subway just past Kenmore Square. It opened for service on July 4, 1959 as the 'Riverside Line'.

At Brookline Jct. on April 19, 1958 there is an inbound B&A *Beeliner* on the main, while the Newton Highland Branch tracks come in from the left. Yes, Red Sox fans, those are the lights of Fenway Park in the background.

Ten years earlier, Sunday afternoon, September 12, 1948, finds a local on the fill near Waban behind 2-6-6T #307.

THE SALVATION ARMY
SOCIAL SERVICE CENTER
FOR THE REHABILITATION OF MEN

Alco RS-3 #8307 with its local passes Crystal Lake on the Newton Highland Branch May 27, 1958, only three days before the end of B&A service.

A local pauses at Elliot behind Alco RS-3 #8249 on April 19, 1958.

In 1949 ALCO-GE commissioned Howard Fogg to create the painting for this advertisement featuring RS-3 #8227 in B&A commuter service.

RS-3 #8305 rounds the curve by Riverside station on May 23, 1958. Soon it will head inbound on the main toward Boston.
Photo from Lawson's collection by Norton D. (Skip) Clark

Trackside from Boston's Suburbs to Springfield

The end of the Boston commuter district was Framingham, 21 miles west of South Station. It also was the LIMITEDs' first stop west of Newtonville. On September 28, 1963 the westbound NEW ENGLAND STATES is still a respectable train behind E8 #4038, an E7B, and trailing E8.

Some six years later on July 8, 1969, Pennsylvania U-30C #6538 and other power rolls a freight westbound through Framingham. The PRR keystone is as red as the ink on Penn Central's ledger. This is the newest action shot in this book.

Lawson visited Webster Jct., a few miles west of Worcester, on June 6, 1949. B&A H5j class Mike #1204 is setting off a car with the semaphore home signal's red board behind her.

Train #11, The Boston Section of the SOUTHWESTERN LIMITED rolls through Webster Jct. at speed behind L3b Mohawk #3052. The westbound grade up Charlton Hill is dead ahead.

Pacemaker Freight

between NEW ENGLAND and NEW YORK and the MIDDLE WEST

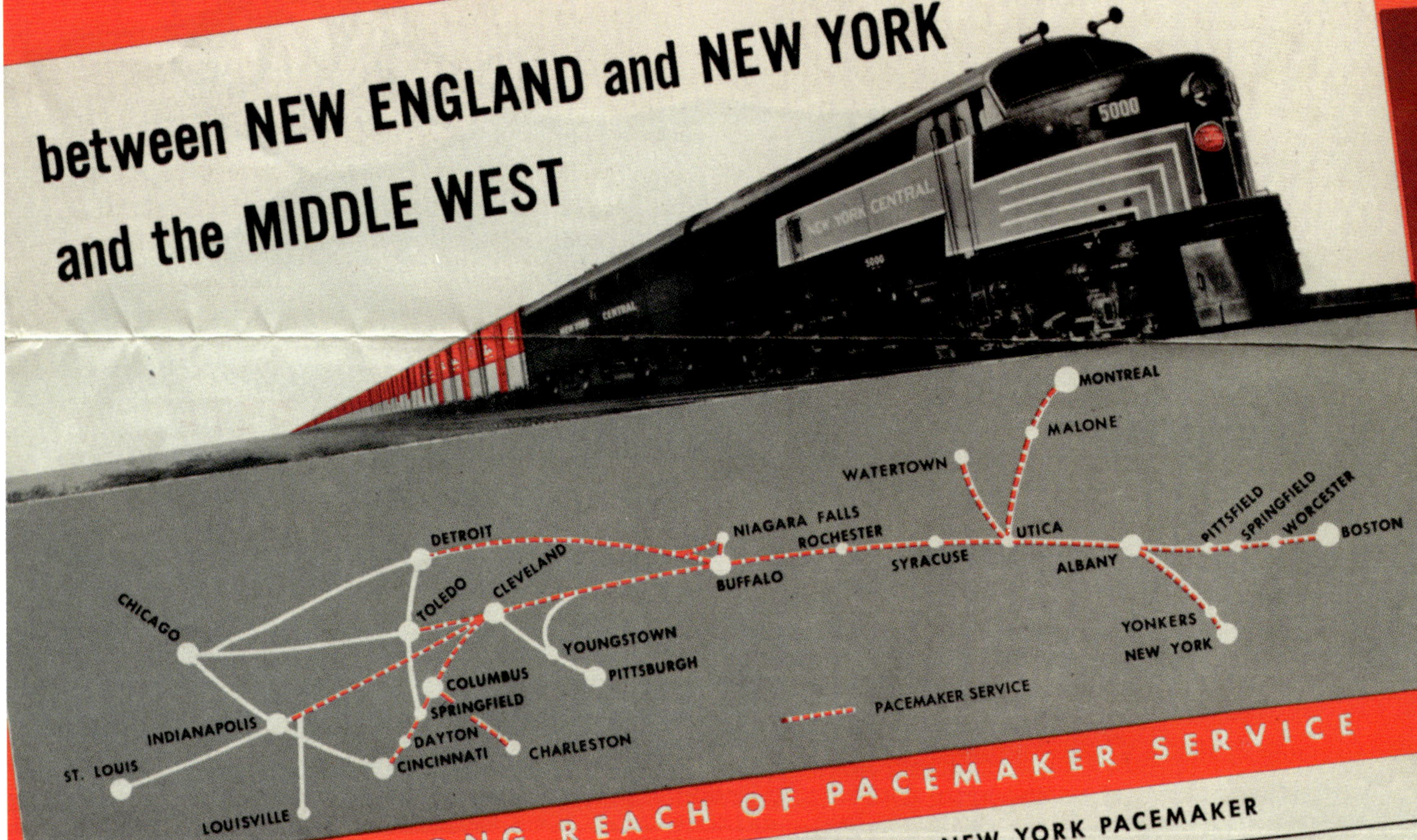

THE NEW LONG REACH OF PACEMAKER SERVICE

NEW ENGLAND PACEMAKER

Westbound • Read Down Prevailing Time	Example by Day	CITIES SERVED		Eastbound • Read Up Prevailing Time	Example by Day
3:40 pm	Mon.	Lv. Boston (Beacon Park)	Ar.	9:25 am	Wed.
5:10 pm	"	Lv. Worcester, Mass.	Ar.	8:25 am	"
7:15 pm	"	Lv. Springfield, Mass.	Ar.	6:50 am	"
8:50 pm	"	Lv. Pittsfield, Mass.	Ar.	5:20 am	"
10:00 pm		Ar. Albany, N.Y. (Rensselaer)	Lv.	4:00 am	Wed.
7:20 am	Tues.	Ar. New York City (33rd St.)	Lv.	7:15 pm	Tues.
2:00 am	"	Ar. Utica, N.Y.	Lv.	1:30 am	Wed.
3:30 am	"	Ar. Syracuse, N.Y.	Lv.	11:20 pm	Tues.
5:20 am	"	Ar. Rochester, N.Y.	Lv.	9:50 pm	"
7:00 am	"	Ar. Buffalo, N.Y.	Lv.	7:50 pm	"
...		Ar. Niagara Falls, N.Y.	Lv.	5:25 pm	"
3:00 am	Wed.	Ar. Detroit, Mich. (Windsor)	Lv.	3:40 am	"
1:45 pm	Tues.	Ar. Cleveland, Ohio (Collinwood)	Lv.	8:15 am	"
11:00 pm	"	Ar. Toledo, Ohio	Lv.	12:30 am	Tues.
5:00 am	Wed.	Ar. Indianapolis, Ind.	Lv.	6:30 pm	Mon.
5:30 am	"	Ar. Cincinnati, Ohio	Lv.	5:15 pm	Mon.

NEW YORK PACEMAKER

Westbound • Read Down Prevailing Time	Example by Day	CITIES SERVED		Eastbound • Read Up Prevailing Time	Example by Day
7:15 pm	Mon.	Lv. New York City (33rd St.)	Ar.	7:20 am	Thur.
10:50 pm	"	Ar. Albany, N.Y. (Rensselaer)	Lv.	3:50 am	"
12:01 am	Tues.	Lv. Albany, N.Y. (Rensselaer)	Ar.	3:30 am	"
2:00 am	"	Ar. Utica, N.Y.	Lv.	1:30 am	Thur.
...		Ar. Malone, N.Y.	Lv.	7:15 pm	Wed.
1:00 am	Wed.	Ar. Montreal, P.Q.	Lv.	...	
2:30 pm	Tues.	Ar. Watertown, N.Y.	Lv.	6:15 am	Wed.
2:30 am	Tues.	Lv. Utica, N.Y.	Ar.	12:55 am	Thur.
3:00 am	"	Ar. Syracuse, N.Y.	Lv.	11:55 pm	Wed.
3:20 am	"	Lv. Syracuse, N.Y.	Ar.	11:35 pm	"
5:20 am	"	Ar. Rochester, N.Y.	Lv.	9:50 pm	"
5:35 am	"	Lv. Rochester, N.Y.	Ar.	9:35 pm	"
6:55 am	"	Ar. Buffalo, N.Y.	Lv.	8:05 pm	"
8:25 am		Ar. Niagara Falls, N.Y.	Lv.	5:25 pm	"
3:00 am	Wed.	Ar. Detroit, Mich. (Windsor)	Lv.	3:40 am	"
1:45 pm	Tues.	Ar. Cleveland, Ohio (Collinwood)	Lv.	8:15 am	We
11:00 pm	"	Ar. Toledo, Ohio	Lv.	12:30 am	T
5:00 am	Wed.	Ar. Indianapolis, Ind.	Lv.	6:30 pm	M
3:15 pm	"	Ar. Louisville, Ky.	Lv.	7:30 pm	M
11:20 am	Thur.	Ar. Charleston, W. Va.	Lv.	11:30 pm	T
7:50 am	Wed.	Ar. Columbus, Ohio	Lv.	7:30 pm	
3:45 am	"	Ar. Springfield, Ohio	Lv.	8:30 pm	
4:45 am	"	Ar. Dayton, Ohio	Lv.	7:30 pm	
5:30 am		Ar. Cincinnati, Ohio	Lv.	5:15 pm	

ADDITIONAL MERCHANDISE SERVICES

For information on other LCL merchandise cars regularly operated between principal points on the New York Central System and to off-line transfers, please consult your local freight traffic representative.

Your *Pacemaker* Merchandise Service is as near as your telephone...

Call any of the friendly, helpful New York Central representatives listed below to learn exactly what *Pacemaker Freight* service can mean to you. He'll gladly give you full information about pickups and deliveries . . . schedules between your community and your shipping destinations . . . less than carload rates on your particular type of merchandise. He will even arrange for an overall study of your shipping needs . . . to help you plan savings of time and money through *Pacemaker Freight*. His expert help is as near as your telephone . . . and it's yours for the asking!

West of Worcester at Milepost 50, back-to-back Alco FA cabs lead a scheduled *Pacemaker* freight west on June 11, 1949. Note the original paint scheme having only a partial side stripe as on the FM 'Erie-builts'.

Printed in U. S. A.
R-6952-4/49-11M

BOSTON & ALBANY

NEW YORK CENTRAL SYSTEM

B&A J2a Hudson #600 on First #559 roars through Webster Jct. on June 11, 1949. It's the late afternoon Springfield local carrying through cars to New York City via Hartford and New Haven.

The Springfield Line

B.&A.-N.Y.N.H.&H.

THROUGH TRAIN SERVICE

BETWEEN

BOSTON

HARTFORD

NEW HAVEN

AND

NEW YORK

BOSTON AND ALBANY RAILROAD

EFFECTIVE APRIL 25, 1937

Matched ABA F3 set NYC #3500-3600-3501 roll train #98 - THE PAUL REVERE - east through Rochdale on October 12, 1947.

Twin Alco road switchers head Train #11 at Rochdale, a few miles west of Webster Jct. on May 22, 1955.

The hill at Charlton, MA. is steepest against westbound traffic. Lawson visited on March 25, 1950 and left us the following five images.

With F3A #3501 leading, the 4500 HP ABA set of EMD locomotives rolls Train #46 up Charlton toward Worcester and Boston. Albany departure was an early 3:45 AM.

Next comes the STATES. Train #28 attacks the eastbound grade behind a matched ABA set of EMD F3's led by #3503.

Train #48, the eastbound Springfield-Boston local, rolls behind a B&A J2 class Hudson in full stride. Leaving Springfield ten minutes behind #28, it provided local service for those in towns not served by the NEW ENGLAND STATES.

In a glorious show of smoke and steam, Train #6, the all-stops local from Albany, thunders upgrade behind yet another NYC Mohawk. It took 7 and 3/4 hours for the trip.

Led by #1005, an ABA set of Alco FA-1 cabs wearing their as-delivered NYC paint scheme grind up Charlton Hill. Their tractive effort is greater, but the FA's passage is much less magnificent than that of the Mohawk which preceded them.

On a colorful October 12, 1947, B&A Pacific #580 rolls Springfield local #551 - with through New York City cars via Hartford - near the crest of Charlton hill.

Moving further west near South Spencer, Lawson found eastbound Train #550 from Springfield rolling through the fall foliage on Columbus Day, 1947.

The westbound NEW ENGLAND STATES behind a B&A Hudson drops downgrade approaching South Spencer. Steam and New England 'Color' are a beautiful combination.

At South Spencer on September 21, 1950, NYC Alco RS-3 road switchers #8225 & 8226 pass with Train #49, the Boston Section of the ADVANCED KNICKERBOCKER.

The eastbound NEW ENGLAND STATES rolls beside a lake at East Brookfield on May 22, 1955. Led by an E7-E8 duo, she makes a grand sight indeed.

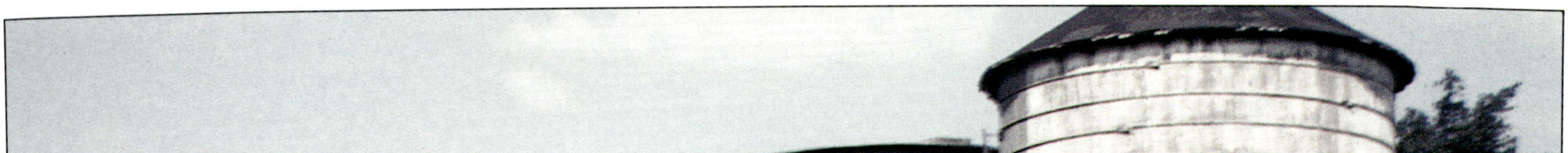

Train #6 behind L3a Mohawk #3024 pauses for a drink from the twin water tanks at West Brookfield on October 2, 1948. With the pops lifting she's ready for the climb ahead. These 4-8-2's were better suited to B&A grades than the J2 4-6-4 Hudsons.

On September 17, 1950 Train #98, THE PAUL REVERE passes through West Brookfield behind Alco RS-3 road switchers #8229 and -30. The effort to keep them in sequential sets will not last much longer. Since 1948 one of the twin water tanks has been removed.

With the verdant green of New England late summer foliage as background, NYC L3a Mohawk #3024 rolls the STATES through West Springfield. The grades of the Berkshire Hills await the passage of the Pride of the B&A this September 17, 1950.

Extra #1023 behind an ABA set of Alco FA-1's drops downgrade at Woronoco. The B unit has been repainted with a side stripe, but both A units still have only partial ones. The coupler cover is still in place on #1023. Just a little fall color is showing beside the Westfield River this September 17, 1950.

Trackside over the Berkshire Hills: The B&A Triumphant

Over 100 miles west of Boston proper is undoubtedly the most famous portion of the B&A, its climb over the mountainous Berkshire Hills of Western Massachusetts. Here, in 1925 Lima Locomotive's first Super-Power 2-8-4 was put to the test, and set new standards for freight haulage at speed and up gradients. By 1930 the B&A owned 55 of the 1400-Class 2-8-4 locomotives.

The 2-8-4 locomotive design was henceforth known in the industry as the Berkshire type. (Except on competitor Boston & Maine, their 25 were known as Limas. No quarter was given in the B&A/B&M Berkshire railroading rivalry). Here too, in the post-war years were the diesel-electric products of Alco, Baldwin, Fairbanks-Morse, Lima-Hamilton, and Electro-Motive tested. Suffice it to say that on April 16, 1951 the B&A was all-diesel.

The Berkshire Hills remain a testing ground for new locomotives. In the mid-60's imported diesel-hydraulics were tested here, and in the late 90's B&A successor Conrail introduced its SD-80MAC's here. These 5000HP V-20 six axle monsters are GM/EMD's most powerful single 2-cycle engine locomotives ever built and use new AC technology.

Another eastbound at Woronoco, this time behind an ABA set of Fairbanks-Morse cab units grind down the valley in the afternoon sun. The FM set is now in the standard paint scheme with a full side stripe. Those three opposed-piston engines are throwing the exhaust high while restraining the train in dynamic brake mode.

With shadows long, the NEW ENGLAND STATES behind an E7 AB set and Mohawk 4-8-2 lead helper attack the Berkshire grade. Note the three smooth-sided cars in the consist this September 17, 1950.

The following day, September 18th, eastbound Train #78 passes through Russell. Lead unit #3203 and sisters were powered with supercharged inline-six cylinder Baldwin powerplants. New York Central had two 4500 HP sets; #3200/3300/3201 and #3202/3301/3203 - shown here. They were 3/4ths of Baldwin's total production of this model, and quickly earned the nickname, 'Gravel Gerties'. In later years they were re-powered with EMD 567 engines.

Later that same day, elephant-eared L3a Mohawk #3022 has Train #98 THE PAUL REVERE, rolling down the right-of-way. Passengers are undoubtedly enjoying that famous Berkshire foliage.

With a shallow Westfield River in the foreground, Train #49, the Boston Section of the ADVANCED KNICKERBOCKER ascends through Russell behind a laboring Mohawk. Only a single streamlined coach interrupts the otherwise heavyweight consist this May 30, 1949.

Alco RS-3 road switchers #8225 & 8226 roll upgrade with four-car Albany local #575 on September 18, 1950.

An Alco FA-1 ABA set behind #1017 shows off its as-delivered NYC paint scheme of steam locomotive black FB booster and only partial side stripes on both A units. In these early years, even the coupler cover is still in place! The location is between Huntington and Chester.

With fall color developing this October 2, 1948, a 4-8-2 has the heavyweight STATES in hand between Huntington and Chester.

Chester was the site of a B&A pusher roundhouse and servicing area. On September 20, 1947 H5j Mikado 2-8-2 #1211 gets a drink at the standpipe before the next push west.

There's a single streamlined sleeper evident in today's Boston Section of the ADVANCED KNICKERBOCKER between Chester and Middlefield.

Another day, another Mohawk. Train #575 westbound to Albany rolls through glorious fall color between Chester and Middlefield on October 2, 1948. Soon enough both glories will pass away.

The following day, October 3rd, at Becket, FA #1005, FB #2303, and FA #1004 sit on the eastbound main amidst the colorful foliage. The train is stopped, and the crew gladly poses for the photographer. The coupler shrouds are open - but still attached.

Train #6 is leaving Becket eastbound behind another Mohawk this September 21, 1950.

Rolling across a high fill with one of the Whistler-built 1841 stone arch bridges ahead over the Westfield River, an ABBA set of FA's is putting its combined 6000HP to good use. Note the sand going down under the wheels to boost traction. This shot was taken from the front of an RRE excursion Budd RDC on April 28, 1957.

NYC *Beeliner* M-460 at the Twin Ledges on April 28, 1957. The fans are out exploring.

One of the ten stone arch bridges of the original 1841 B&A alignment built under the supervision of Chief Engineer George Washington Whistler. Eight were eliminated with a 1912 right-of-way realignment, but two remain in service over a century and a half after construction. Several - like this example - remain standing today as a symbol of his engineering prowess. Steps are now (2000) being taken by the Mass. Dept. of Environmental Management to preserve these arches for future generations to admire.

L3a Mohawk #3002 with Train #78 from Albany is running through the fall color between Dalton and Hinsdale on October 3, 1948.

At Dalton, a Mohawk is but a blur as she leads an ABA set of EMD F3's upgrade on September 21, 1950. Train #28, the eastbound NEW ENGLAND STATES, has more than sufficient power today.

On April 28, 1957 we get another view of action in Dalton. This time westbound #27 - from the cab of an eastbound *Beeliner* carrying the RRE excursion.

That April 28, 1957 RRE Budd RDC excursion left the B&A main at North Adams Jct. - just east of Pittsfield - and continued north to the City of North Adams. Here NYC *Beeliner* M460 poses in the North Adams yard.

At West Pittsfield on September 18, 1950, we view our last B&A steam image. Train #49, the ADVANCED KNICKERBOCKER, is leaving the Berkshires behind in a cacophony of roaring steam and flashing steel thanks to L3a Mohawk #3002.

State Line Tunnel marks the end of the B&A over the Berkshires. Located just over the border in New York, the two bores provide an appropriate spot for us to complete our B&A visit. Unchanged for decades, the picture of the tunnels was taken on May 2, 1964. In 2000, the left-hand bore is trackless.

New York Central and the E7

The most successful diesel-electric passenger locomotive built in post-war America - with 511 units - was the Electro-Motive E7. Containing twin V-12 model 567 engines, and riding on two A-1-A trucks, the streamlined 71' long units had an undeniable Art Deco majesty. An A-B combination netted 4000 horsepower, and provided starting - but not high speed - tractive effort well in excess of a 4-8-4 Niagara, much less a Mohawk or Hudson.

When A-B or A-B-A sets were permanently assigned to the prestigious 20TH CENTURY LIMITED, a tidal wave of change to New York Central System's motive power was confirmed.

New York Central ultimately fielded a fleet of 37 E7A and 14 E7B units. Of the 480 successor 2250 horsepower E8 models built, NYC purchased 60 E8A's. The E7's lasted through the Penn Central era with some E8's going to Amtrak in 1971. PC successor Conrail operated its Business Train behind E8s, which now belongs to successors CSX and NS.

In the words of the Central's 1940s-1960s company photographer - Ed Nowak - the B&A NEW ENGLAND STATES LIMITED was considered second only to the CENTURY in panache. When members of that 1945 group of eight A and four B E7 units appeared in the late Spring of 1947 on #27 and #28, Lawson took notice. Following is a vignette of E units from 1947-1971 powering the B&A's flagship. All but the final image are taken at Newtonville.

NYC #4000 and a sister E7 are clad in the as-delivered paint scheme of a dark gray body and light gray stripe on October 26, 1947. Note that the multi-striping over the nose is higher on these units, enclosing the original style number boards. This scheme coordinated with that on the 1938 CENTURY and its streamlined J3a Hudson power. As early as 1940, EMC proposed their model E6 to the NYC, as a color illustration of one in this garb exists. The Hudsons prevailed - for a while.

The as-delivered, but experimental, all black scheme found only on E7's #4002 & #4003. This scheme appears to be based upon NYC's wartime one on FT's #1600-03A/2400-03B delivered in 1944. Once again, the STATES has new power this June 1, 1947.

A member of the second batch of E7's - #4014 - glides through town on May 17, 1947. The now standard grey with light grey trim paint scheme matches that on the abuilding 1948 CENTURY.

NYC #4004 A and matching B-unit pause with the outbound STATES on May 21, 1949. They still carry their as-delivered paint scheme. Note the side stripe zigzag on the B-unit. As delivered, the A-B sets were considered matched sets and even the B-units had a front and back.

With over 15 years of service, E7 #4019 still looks well kept, along with E8 #4025 and an unidentified 2nd E8 leading #27 on June 3, 1962. Times are changing, and the two closer tracks have been removed as part of the construction of the Massachusetts Turnpike Extension from circumferential highway Mass Route #128 into downtown Boston.

The successor of the E7, the 2250 horsepower E8, was delivered exclusively in the two-tone grey scheme standardized on the later E7s. Seeking a more economical and new image, the conservative NYC did not do itself proud with this green scheme in the early 1960's. Fortunately its error was quickly realized, and only three E8's were so painted. E8 #4083 awaits the highball on an undoubtedly truncated NEW ENGLAND STATES.

The final indignity. PC E8 #4058 and snackbar coach #3204 are the tattered remains of the NEW ENGLAND STATES. This April 8, 1971 image of eastbound PC Train #428 at Framingham is a far cry from the chic image and promise of the lounge, diner and sleepers of #27 and #28 in 1949. Yet the E-unit retains its classic lines.

For an interesting comparison with B&M E7s, the reader is directed to ***Boston & Maine Trackside with Arthur E. Mitchell*** pages 36-41.

the NEW YORK, NEW HAVEN & HARTFORD RAILROAD

AS THE LARGEST NEW ENGLAND RAILROAD, IT IS LOGICAL TO ANTICIPATE that the New Haven section of ***Trackside* around Boston** would be the book's largest. However, Lawson Hill apparently sold off much of his New Haven material during his later years. We therefore are rather limited in number of images, but certainly not in quality of content.

The greatest concentration of New England's manufacturing facilities, educational institutions, financial houses and raw political power was located within the New Haven's territory. Even including financial scandals and moral lapses, the New Haven rode well its peak of power, influence and monopoly across the first half of the 20th Century. The road faltered as heavy industry withered and passengers defected to the turnpikes and air shuttle in the 1950's. New Haven entered bankruptcy in 1961. It was absorbed into Penn Central on January 1, 1969, and became part of Conrail in 1976. Since the 1980's a rebirth of its passenger services has occurred under Amtrak and the various regional transit authorities.

Our visit begins at South Station, since 1899 the road's Boston terminus. South Station was actually operated by the Boston Terminal Company; a jointly owned subsidiary of the New Haven, Old Colony (controlled by the New Haven), and Boston & Albany.

South Station's center tracks were the longest, and generally handled the great Limiteds of both the B&A and New Haven. Service to the Cape and Southeastern Massachusetts - over the 'Old Colony' lines - proceeded straight south over the Fort Point Channel over three lift bridges. Local and suburban service down the Midland Division - the original 'New York & New England' via Willimantic and Hartford - also left this way.

The New Haven's legendary Shore Line Service to New York City and Washington bore to the right of the Fort Point Channel and paralleled the B&A tracks west to Back Bay Station. The Shore Line swung south away from the B&A just past Back Bay, and headed toward Providence. At Readville the Shore Line passed under the east/west Midland Division. Readville was also home to a large New Haven shop complex until closing in 1959.

The New York New Haven and Hartford RAILROAD CO.

The B&A generally operated at the western side of South Station, with all service flowing out its main west to Trinity Place or in from Huntington station in the Back Bay. Its freight houses were located on the western side of the station and out along the main for some distance.

To the south of the station lead tracks, New Haven freight yard leads stretched into large yards in South Boston for maritime service. The passenger yards plus Southampton Street and Dover Street engine terminals were also in this same general area.

Although frequent changes were made to this general layout throughout the years, it does give a brief overview for those who have little knowledge of the general South Station and associated yard areas.

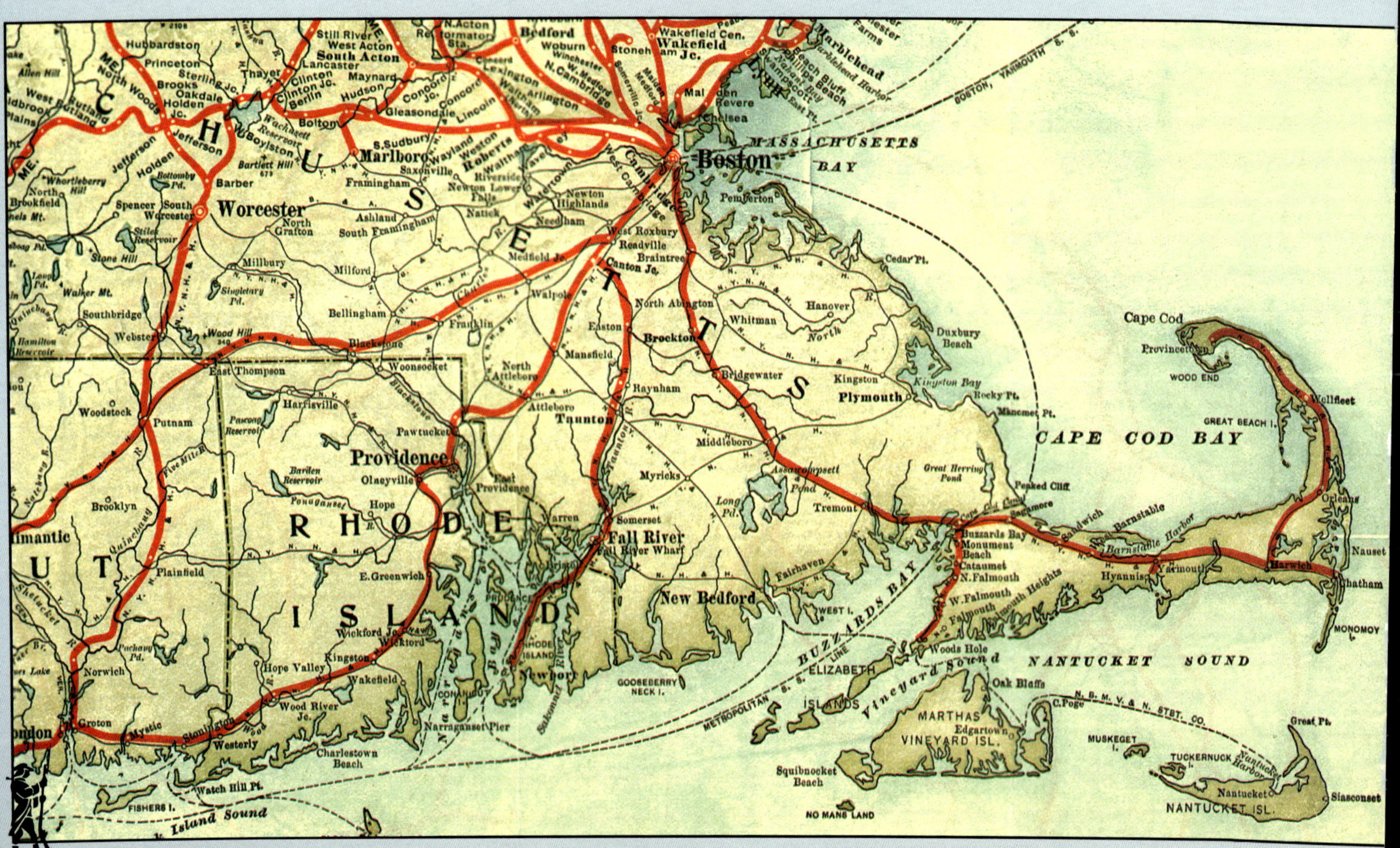

New Haven Steam Locomotive Roster Circa 1945-1952 (End of steam).

TYPE	CLASS	NUMBER	DATE(S) BUILT	BUILDER
0-6-0	T-1/-2a	#2300-2343	1900-02/1904	RI/Schenectady
	T-3	#2350-2355	1904-1910	Schenectady
	T-2b	#2373-2469	1905-1913	Cooke/RI/Schn.
0-8-0	Y-3	#3400-3434	1920-1923	Alco: Schenectady
	Y-4/-4a	#3600-3615	1924/1927	Alco: Schenectady
2-6-0	K-1b/-1d	#260-279 &	1900-1907	Schen/RI/Cooke
		#325-479		BLW/Schenectady
2-8-0	F-5	#150-164	1912	Alco: Schenectady
2-8-2	J-1	#3000-3024	1916	Alco: Schenectady
	J-2	#3100-3107	1916	Alco: Schenectady
2-10-2	L-1	#3200-3249	1918	Alco: Schenectady
4-4-2	H-1	#1100-1111	1907	Alco: Schenectady
4-6-0	G-4a	#800-849	1904-1907	BLW
4-6-2	I-1	#1000-1031	1907/1910	Alco: Schen & BLW
	I-2	#1300-1349	1913	Alco: Brooks
	I-3	#1090-1095	1913	BLW
	I-4	#1350-1399	1916	Alco: Schenectady
4-6-4	I-5	#1400-1409	1937	BLW
4-8-2	R-1/1a/b	#3300-3348	1919-20/1924	Alco: Rich. & Schen.
	R-2/2a	#3500-3507	1924/1926	Alco: Schenectady
	R-3/3a	#3550-3562	1926/1928	Alco: Schenectady

New Haven all-time Electric Locomotive Roster

Experimentals	none	#068/069-072	1910-1912	Alco-GE/BLW-West.
Passenger:				
B+B	EP-1	#01-041	1906-1908	BLW-Westinghouse
1-B+B-1	EP-1 All modified 1908-1909			
1-C-1+1-C-1	EP-2	#0300-0326	1919-1928	BLW-Westinghouse
2-C+C-2	EP-3	#0351-0360	1931	General Electric
2-C+C-2	EP-4	#0361-0366	1938	General Electric
C-C	EP-5	#370-379	1955	General Electric
Yard:				
B+B	EY-2/-2a	#0200-0215	1911-1912	BLW-Westinghouse
	EY-2b	#0218-0223	1926-1927	BLW-Westinghouse
	EY-2c	#0224	1911	BLW-Westinghouse
B+B	EY-3	#0216-0217	1926	Alco-GE
Freight:				
1-B+B-1	EF-1	#073-0111	1912-1913	BLW-Westinghouse
1-B+B-1	EF-2	#0112-0116	1926	Alco-GE
2-C+C-2	EF-3	#0150-0154	1942	BLW-Westinghouse
	EF-3a	#0155-0159	1943	General Electric
C-C	EF-4 (xVGN)	#300-310 (2nd)	1956-1957	General Electric

Note: All Zero (0) prefixes removed from electrics by April 15, 1950.

New Haven all-time Diesel-Electric Locomotive Roster*

NUMBER	BUILDER	TYPE	YEAR BUILT
#0400-0429	Alco/GE	FA-1	1947
#0450-0464		FB-1	1947
#465-469		FB-2	1951
#0500-0516	Alco/GE	RS-2	1947-1948
#517-561		RS-3	1950-1952
#590-599	F-M	H16-44	1950
#0600-0621	Alco/GE	S2	1944
#0630-0639	L-H	1200 HP Switcher	1950
#640-659	EMD	SW-1200	1956
#0660-0671	Alco/GE	RS-1	1948
#0700-0759	Alco/GE	DL-109	1941-1945
#0760-0786		PA-1	1948-1949
#790-799	F-M	CPA-24-5 (#790-91 ex F-M demos)	1950/1952
#0800-0818	GE	44-Tonner.	1941-1947
#0900	Alco/GE	None; 1st diesel switcher in N. Eng.	1931
#0901-0905	C-B/GE	None; built w/C-B engines	1936-1937
#0906-0910	I-R/GE	None; built w/I-R engines	1936
#0911-0920	Alco/GE	HH-600	1938
#0921-1930	Alco/GE	HH-660	1939-1941
#0931-0995	Alco/GE	S-1	1941-1949
#1200-1229	EMD	GP-9	1956
#1400-1414	Alco	DL 701	1956
#1600-1614	F-M	H16-44	1956
#2000-2029	EMD	FL-9	1956-57
#2030-2059	EMD	FL-9 (#2059 last EMD F-cab built)	1960
#2500-2525	GE	U-25B	1964-1965
#2550-2559	Alco	C-425	1964-1965
#3000-3001*	B-L-H	RP-210* (*Dan'l Webster* locomotive) *diesel-hydraulic	1956
#3100-3101	F-M	P12-42 (*John Q. Adams* locomotive)	1957

Trackside Boston to New Haven via the Shore Line

South Station's main entrance faces Dewey Square, the junction of Summer Street and Atlantic Avenue. Here, on August 26, 1956 the steel girders covering the Dewey Square tunnel - part of the new Southeast Expressway - are freshly installed. This change to underground construction was wrought by massive public protest at the previously constructed elevated expressway section by North Station. Its completion contributed towards the end of all Old Colony south shore service on June 30, 1959.

The view down South Station platform tracks on June 5, 1949. Note the early New Haven 'High Hood' Alco switcher.

A mighty I-5 streamlined Hudson rests at Southampton St. engine terminal on March 6, 1948. Delivered by Baldwin in 1937, #1401 and her nine sister I-5's were designed to handle a 12 car, 830 ton Shore Line passenger train at 60+ MPH up the .7% grade of Sharon Hill. They did, but bent the rails in the process, until rebalanced by the New Haven mechanical department in conjunction with Baldwin.

Also at Southampton Street that same March day, three-cylinder Mountain type class R3a #3557 awaits its orders with just a whiff of steam escaping from the turbogenerator. The 4-8-2 came from Alco in 1928.

On the Old Colony - now a commuter line - a local behind I-2 Pacific #1306 rolls over a wood pile trestle between stops at Atlantic and Neponset on October 4, 1947.

At Readville, where the Midland Division to Putnam and Hartford, CT. leaves the main, three cylinder 0-8-0 switcher #3613 stands in the snow on March 6, 1948. Note that the valve linkage to the inside cylinder is visible in front of the cylinder on the engineer's side.

One of the rarest images in this collection. Nearly new PA's #0762 and #0763 crest Sharon Hill with the New York City-bound PATRIOT on Independence Day, 1949. The 'Brooks Brothers' paint scheme of warm orange with 5 silver-gray pinstripes didn't last long - especially clean. This slide has been rescued from Ansco Red thanks to computer color correction.

Lawson visited the Charles Street engine facility in Providence on August 31, 1947. F5, or 'Bull Moose' class Consolidation #150 originally was a Central New England engine.

I-1 Pacific #1005 stands beside the Charles Street coal dock in Providence.

Further away from the coal tower - for which they have no need - Alco S-1 switchers #0939 and #0933 burble away.

RS-2 #0503 is coupled to an unidentified Alco switcher at Charles Street on June 5, 1949.

RDC-3 #125 and an unidentified RDC-1 depart New London for Boston via the Shore Line on a dreary May 16, 1954.

New Haven Mike #3013 with white trim on driver tires and engine catwalk pauses at New Haven - note the overhead catenary. The fresh appearance and white flags signify that this is the New Haven steam fantrip of June 5, 1949. That 'American Flyer' coach appears sparkling clean as well. These cars, built by Osgood Bradley between 1935-38 were nicknamed so because they were copied by A. C. Gilbert's American Flyer Trains in O- and S-Gauge by the thousands between 1941 and 1954.

The New Haven's Alco DL-109 Passenger/Freight Locomotives

The New Haven purchased New England's first diesel-electric locomotive in 1931, a 600 HP one-of-a-kind Alco switcher containing a McIntosh & Seymour (Alco) Model 531 in-line six cylinder engine. Impressed, the New Haven rostered 20 similar 'High Hood' Alco switchers by WWII.

With diesel efficiency and availability duly noted, the New Haven was immediately interested in Alco's new diesel-electric passenger locomotive, a direct competitor to the GM/EMC E-unit. An initial-production prototype built for the Rock Island was completed on January 3, 1940. The next year the New Haven ordered ten model DL-109 A cabs, planned for back-to-back operation on both its fastest Shore Line Limiteds and scheduled freighters between Boston and New Haven. With that dual service operation in mind, Alco was instructed to beef up the locomotive's draft gear and nose end framing.

The DL-109 achieved its 2000HP from two turbocharged model 539T in-line six cylinder engines. This powerplant was an upgraded and turbocharged version of the model 531 contained in the earlier Alco switchers - there would be no surprises for the maintenance department. Likewise with the General Electric traction motors and electrical gear. The first unit, NH #0700, was shipped from Alco on December 5, 1941, and three more followed before month's end. All 10 locomotives were on the property by April 13, 1942.

The first use of the DL-109's was with #0700 and #0701 leading the southbound COLONIAL on December 13, 1941, six days into WWII. With War Production Board controls implemented on April 4, 1942, additional units were slow to arrive.

Soon the DL-109's were working Shore Line passenger jobs during the day and evenings, then handling freight between New Haven and Boston during the wee hours. The 80 MPH geared units were so productive that the nearly new (1937) I-5 semi-streamlined Baldwin Hudsons became secondary power on the now-stretched-to-capacity Limiteds.

Units #0710-19 were delivered between July, 1942 and January, 1943. No other deliveries were made that year. Sisters #0720-29 arrived between January and April, 1944, and #0730-39 came between October, and January, 1945. Units #0740-59 were delivered between January, 1945 and April 13, 1945.

As the number of units increased, and the New Haven became more skilled at locomotive utilization, the 0700's began heading up freights over the Maybrook line via Poughkeepsie, NY, as well. These early first generation diesels performed yeoman service across the New Haven system, and were America's first dual service road locomotives.

In early 1949 the MERCHANTS LIMITED was speeded up to 3 hours and 55 minutes between Boston and New York City. The addition of another 2000 HP locomotive to the matched AA 0760-class Alco PA power was deemed necessary to keep the MERCHANTS on time, so DL-109's #0725 and #0727 became B-units with nose MU cables. NH #0727 even had its nose modified with a passageway door. The assignment was brief, and still operable cab unit #0727 kept its modified nose until scrapping.

An early 1950's proposal from the New Haven's Engineering Department was to convert the aging DL-109 fleet into diesel-electric-electrics with 3rd rail shoes. Before the FL-9's were ordered, they determined what it would require to gut an old DL-109, replace the two engines with one Alco or EMD, and fill the rest of the carbody with straight electrical gear. The project never went past the preliminary engineering stage.

Although the arrival of twenty-seven Alco PA-1 passenger locomotives, thirty FA, and twenty FB freight locomotives in the post-war '40's and 50's cut into their work, the DL-109's still held down main line assignments until the 60-unit road switcher order of 1956 arrived. Those Alco and EMD road switchers, plus the thirty FL-9's delivered in 1957 sent most of the DL-109's to the scrapper, rather than the rebuild shop. The remainder clustered in southeastern Massachusetts' Old Colony local and suburban service.

With the opening of the Southeast Expressway and the termination of Massachusetts subsidy payments for the Old Colony service to the New Haven on June 30, 1959, all Old Colony passenger service was terminated. The last New Haven DL-109 most likely ran on July 2nd, or 3rd with the heavy holiday service to Cape Cod. All were scrapped by 1960 with the exception of #0716. Now called PP (power plant) #716, the tired locomotive provided 600V DC power to test 3rd rail equipment at Readville and Boston. After that shop closed, it lingered in the Dover Street yard until Penn Central cut it up circa 1970.

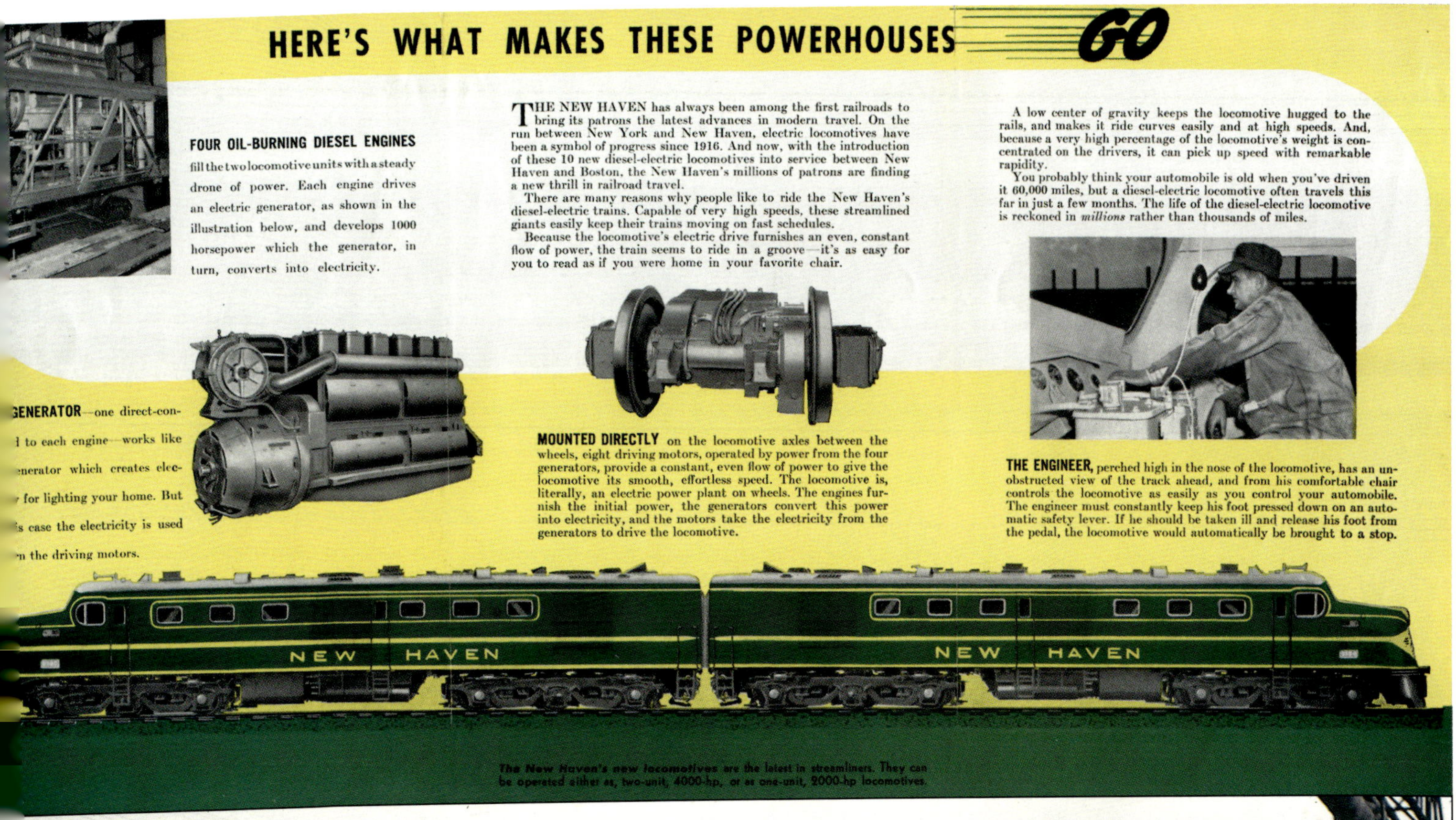

HERE'S WHAT MAKES THESE POWERHOUSES GO

FOUR OIL-BURNING DIESEL ENGINES fill the two locomotive units with a steady drone of power. Each engine drives an electric generator, as shown in the illustration below, and develops 1000 horsepower which the generator, in turn, converts into electricity.

THE NEW HAVEN has always been among the first railroads to bring its patrons the latest advances in modern travel. On the run between New York and New Haven, electric locomotives have been a symbol of progress since 1916. And now, with the introduction of these 10 new diesel-electric locomotives into service between New Haven and Boston, the New Haven's millions of patrons are finding a new thrill in railroad travel.

There are many reasons why people like to ride the New Haven's diesel-electric trains. Capable of very high speeds, these streamlined giants easily keep their trains moving on fast schedules.

Because the locomotive's electric drive furnishes an even, constant flow of power, the train seems to ride in a groove—it's as easy for you to read as if you were home in your favorite chair.

A low center of gravity keeps the locomotive hugged to the rails, and makes it ride curves easily and at high speeds. And, because a very high percentage of the locomotive's weight is concentrated on the drivers, it can pick up speed with remarkable rapidity.

You probably think your automobile is old when you've driven it 60,000 miles, but a diesel-electric locomotive often travels this far in just a few months. The life of the diesel-electric locomotive is reckoned in *millions* rather than thousands of miles.

GENERATOR—one direct-con-
d to each engine—works like
enerator which creates elec-
for lighting your home. But
s case the electricity is used
n the driving motors.

MOUNTED DIRECTLY on the locomotive axles between the wheels, eight driving motors, operated by power from the four generators, provide a constant, even flow of power to give the locomotive its smooth, effortless speed. The locomotive is, literally, an electric power plant on wheels. The engines furnish the initial power, the generators convert this power into electricity, and the motors take the electricity from the generators to drive the locomotive.

THE ENGINEER, perched high in the nose of the locomotive, has an unobstructed view of the track ahead, and from his comfortable chair controls the locomotive as easily as you control your automobile. The engineer must constantly keep his foot pressed down on an automatic safety lever. If he should be taken ill and release his foot from the pedal, the locomotive would automatically be brought to a stop.

The New Haven's new locomotives are the latest in streamliners. They can be operated either as, two-unit, 4000-hp, or as one-unit, 2000-hp locomotives.

Three Alco DL-109's relax at Southampton Street in the sun on April 24, 1948. Engines #0740, #0745, and #0723 each have different paint schemes. In all, the DL-109 fleet had 10 variations - perhaps more - of color schemes. New Haven #0740 was the only DL-109 to retain its original side-window arrangement to the end. The other 59 DL-109's got upper grill screening during shoppings in later years.

The '3PM Train to New York' is picking up speed passing Chickering Tower after departing Back Bay station. The tower was named for the famous Chickering Piano factory located behind it. Two Alco DL-109s head up the heavyweight consist this March 3, 1946.

On May 23, 1948 DL-109 #0736 is passing Chickering, this time with Shore Line Train #15, the PURITAN. Note the partially completed John Hancock (Insurance Company) skyscraper in the background. When opened in 1949, it is New England's tallest building, surpassing the 1915 Custom House tower.

At North Framingham, where the New Haven branch to Concord and Lowell crosses the B&A, Lawson found 0-8-0 switcher #3429 slumbering on June 14, 1949.

Trackside by neighbors B&A and B&M

R1b class Mountain #3341 was also in Framingham that same afternoon. Alco built the 4-8-2 in 1924.

A New Haven RRE excursion arrives in Worcester on June 20, 1948. Power is I-2 Pacific #1343 and four 'American Flyer' cars. The fans are headed for a B&M *Minute Man* F-unit set out of sight to the rear of the special.

B&M F2 #4252 and an FT A-B set rest in New Haven's Southbridge Street (Worcester) yard. Likely the power is awaiting the arrival of Maine-bound tonnage from Cedar Hill yard in New Haven. This yard now serves the Providence & Worcester RR.

The northbound EAST WIND arrives in Worcester behind Alco DL-109 cab units #0710 and #0745 on June 20, 1948. The EAST WIND was MeC/B&M/NH/PRR interline summer daytime service between Bangor, Me., and Washington DC. This season it ran via Providence, with a scheduled northbound arrival in Worcester at 1:30 PM. The EAST WIND started in 1940 with a matched bright yellow and silver trim consist (See pages 18-19 of *Trackside East of the Hudson 1941-1954 with Bill McChesney*). Suspended during WWII the EAST WIND was terminated at the end of the 1955 summer season.

Trackside to Cape Cod

On June 9, 1957 Lawson was on a fan trip to Cape Cod. Budd RDC #44 likely came from Boston via the Old Colony, but perhaps went down the Shore Line to Attleboro and then east to the Cape.

Buzzards Bay is the rail entrance to Cape Cod. At the station NH RDC #44 meets a New York City-bound train - likely the CAPE CODDER - behind DL-109 #0724. A Budd RDC connection to Boston and another one from Woods Hole are also present. Note the RDC reflection in the sparkling 'American Flyer' car side.

A few minutes later NH #0724 departs, with a PRR baggage car and at least two 'American Flyer' cars in tow.

The US Army Corps of Engineers started construction of the Buzzards Bay lift bridge on December 12, 1933, and placed the bridge in operation on December 27, 1935. The Corps has maintained and operated it ever since. The towers are 271' above mean sea level, and the 544' long lift span was for decades the worlds longest. The structure simply dwarfs the Budd RDC on the span. We are on the Cape side of the bridge, with the line east to Hyannis in the foreground, and the branch south to Woods Hole bearing off to the left immediately in front of the bridge.

At its maximum, New Haven track reached out to Provincetown - the absolute end of the Cape, and about 70 miles from the Canal. Here the fans saunter around the Budd RDC in an unidentified location, perhaps near Falmouth.

TRACKSIDE with a HIGH GREEN!

Twin B&A upper quadrant semaphores rule the main near Trinity Place and Back Bay Stations. With its 'High Green' (maximum permissable speed) indication, now is the time to leave *Trackside around Boston*, and all that which was New England Railroading in Lawson Hill's era.

'High Green' still remains over this Boston raill corridor today, as Amtrak's electrified *Acela* Service develops on the Northeast Corridor to New York City and Washington, DC. Increasingly crowded airports and highways bode well for the continuing regrowth of rail patronage to levels unseen since the pre-Connecticut Turnpike era of the New Haven.

Likewise, the Conrail break-up has placed freight service on these tracks in the hands of CSX, one of the two remaining Eastern rail giants. Again, thanks to congestion, additional Amtrak and freight service 'To the West' via Albany seems certain as America grapples with the logistics of the New World economy.

This closing image is both a 'High Green' into Boston's railroad future, and a monument to New England Railroading's glorious past. Our journey with Lawson Hill is complete.